MEI Structured Mathematics

Mathematics is not only a beautiful and exciting subject in its own right but also one that underpins many other branches of learning. It is consequently fundamental to the success of a modern economy.

MEI Structured Mathematics is designed to increase substantially the number of people taking the subject post-GCSE, by making it accessible, interesting and relevant to a wide range of students.

It is a credit accumulation scheme based on 45 hour components which may be taken individually or aggregated to give:

3 components	AS Mathematics
6 components	A Level Mathematics
9 components	A Level Mathematics + AS Further Mathematics
12 components	A Level Mathematics + A Level Further Mathematics

Components may alternatively be combined to give other A or AS certifications (in Statistics, for example) or they may be used to obtain credit towards other types of qualification.

The course is examined by the Oxford and Cambridge Schools Examination Board, with examinations held in January and June each year.

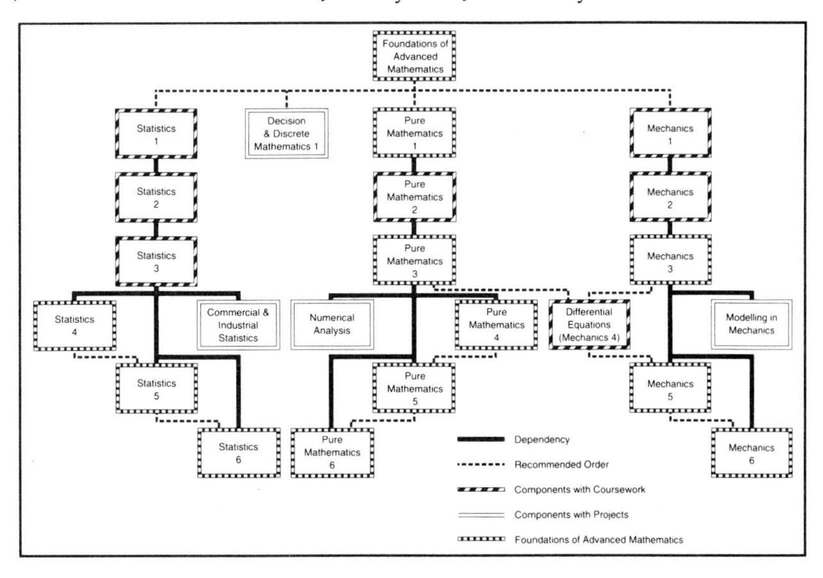

This is one of the series of books written to support the course. Its position within the whole scheme can be seen in the diagram above.

Mathematics in Education and Industry is a curriculum development body which aims to promote the links between Education and Industry in Mathematics and to produce relevant examination and teaching syllabuses and support material. Since its foundation in the 1960s, MEI has provided syllabuses for GCSE (or O Level), Additional Mathematics and A Level.

For more information about MEI Structured Mathematics or other syllabuses and materials, write to MEI Office, Monkton Combe, Bath BA2 7HG.

Introduction

This is the first in a series of books written to support the Mechanics Components in MEI Structured Mathematics, though you may also use them for an independent course in the subject. Throughout the series emphasis is placed on understanding the basic principles of mechanics and the process of modelling the real world, rather than on mere routine calculations.

In this book you meet the basic concepts and laws of motion and force, vector techniques and the modelling cycle. Some examples of everyday applications are covered in the worked examples in the text, many more in the various exercises. Working through these exercises is an important part of learning the subject; not only will it help you to appreciate the wide variety of situations that can be analysed using mathematics, it will also build your confidence in applying it to them.

You should however appreciate that mechanics is not just a pen-and-paper subject. It is about modelling the real world and this involves observing what is going on around you, and when necessary conducting your own experiments. This book includes a number of experiments and investigations for you to carry out. Make sure that you do so; such work will really help you to understand the basics of the subject.

We have used S.I. units throughout this book but have occasionally included an example using other units that you might meet in everyday life (m.p.h., knots and even light years). By convention a vector may be drawn with the arrow marked either at its middle or at its end. In this book, when vectors are being added the arrows are at their middles so that the vector diagrams are easy to read. Otherwise we have marked arrows at the ends of vectors.

Finally we would like to take the opportunity to thank the many people who have helped in the preparation of this book: Chris Compton, Rob Hodges and Nick Pendry who have provided valuable comments on draft material; Sharon Ward for typing the manuscript; and Harry Paticas for the many photographs illustrating the real applications of mechanics.

John Berry, Ted Graham, David Holland and Roger Porkess

Contents

1 Introduction to modelling

The goal of applied mathematics is to understand reality mathematically.

G. G. Hall, 1963

When trying to move a heavy packing case, should you use all your effort to pull it along or would you be better advised to pull partly upwards as well?

When throwing a stone off a cliff, at what angle must you throw it to land in the sea as far away as possible?

When your car is one of a queue stopped at traffic lights and the lights turn green, how much time elapses before you drive past them?

If you think for a moment about what you do on a typical day, many other questions about forces and motion will readily come to your mind.

In this book you begin to learn how to analyse such problems. The area of applied mathematics which studies the motion of objects is called *mechanics*.

Most problems in the real world seem more and more complicated the longer you think about them. Take the problem about the packing case. If you use part of your effort in lifting, it may be easier to slide, but if you have less effort available to pull it have you gained anything? Will the packing case be deformed by your efforts? Does it matter? We can go on in this way, considering ever more complicated possibilities until the problem becomes very involved. How should we approach it?

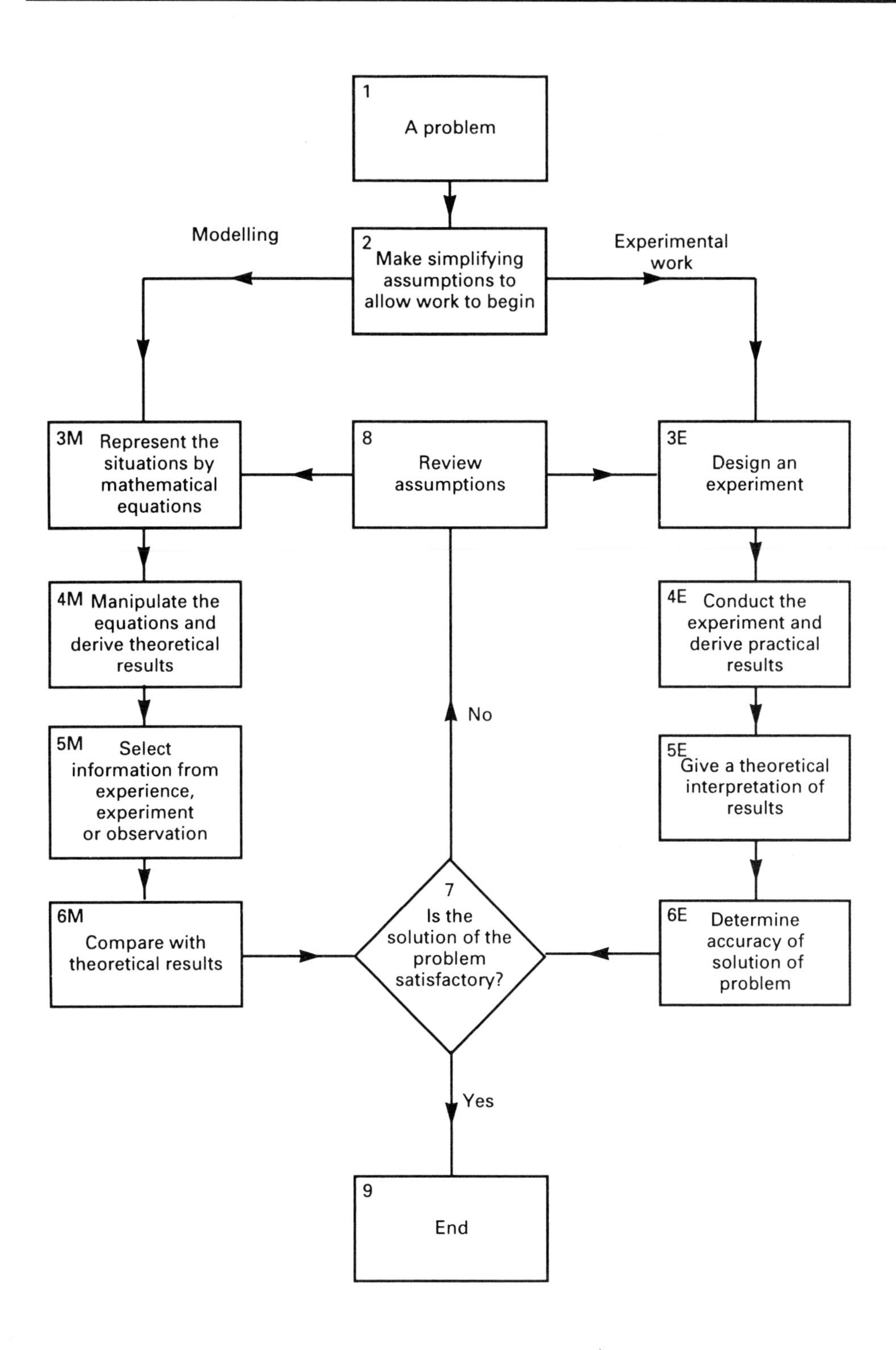

Figure 1.1 Modelling flow chart

The first step is to make simplifying assumptions to allow you to begin work on the problem. You should be quite ruthless about this. Simplify the problem as much as you can, but make sure that you list all the assumptions you make. At a later stage you will need to review them.

Ignore any factors which have no significant effect. These vary from one problem to another, but may perhaps include:

- the size of an object. In mechanics, an object whose size is neglected is called a *particle*;
- the weight of an object. In mechanics, an object with negligible weight is said to be *light*;
- friction. In mechanics, a friction-free surface is said to be *smooth*;
- wind and air resistance;
- spin;
- small variations in quantities involved such as forces or accelerations. Quantities that are constant are said to be *uniform*.

Remember that you will probably go through the problem several times, progressively removing these assumptions.

You should by now be able to start work on the problem. (If not, perhaps you have not simplified it enough.) Begin by using very easy numbers before taking more complicated ones or looking for the general solution using algebra.

With many problems you will be able to go straight into the mathematics (the left-hand cycle on the flow chart). This will usually involve the sort of work that is in many of the exercise questions later in this book. In these questions you are given the assumptions and you apply the appropriate principles and techniques. It is essential that you master these techniques but you must always remember that their use is only a part of the whole modelling cycle.

In problems of this sort, you are able to express the simplified problem in mathematical symbols and equations, and to use them to work out an answer. Your answer will not necessarily be correct, even if your calculations are sound, because it will depend on the simplifying assumptions you have made. You must therefore find some way of checking your answer. You might have to do an experiment to collect data, or you might be able to test your result against experience or common sense. If there is no way of checking your result against reality, it is virtually worthless.

Often your first result will not be particularly good. You then look at your assumptions and decide which of them you are going to cross out for a second run through the problem. Sometimes you will go through the problem several times, taking out simplifications one by one, until you are satisfied that your result matches reality. It is helpful if your first attempt is almost childishly simple: you know it will have to be refined but at least it allows you to get started.

On the other hand, the first work you do after making your simplifying assumptions may be experimental. This should help you to understand the problem better and give you some figures to work with, but you must then give a mathematical explanation of your results. This approach corresponds to the right hand side of the flowchart.

When you collect data you must recognise their variability, and take enough readings to ensure that your figures are representative of the range of possible values. Never be content with a single reading. You need to be thinking not only of a typical value but of the spread of possible values. Well-known measures of typical values are mean, mode and median; those of spread include range, inter-quartile range and standard deviation.

You also need to ask yourself to what extent the situation involves chance. Take the example of the queue at the traffic lights. Suppose that your car is the fifth vehicle. There is no definite answer as to how long it will take you to reach the lights because it depends on the nature of the vehicles in front. They might be fast cars with alert drivers or heavy lorries that only pull away slowly, or worst of all a slow-moving tractor. It may be appropriate for you to bring probability into your solution of such a problem.

It is essential to estimate the accuracy of your solution to a problem, and to present the answer to an appropriate number of decimal places. Suppose for example that you timed an Intercity train going past you, your stop-watch giving a reading of 4.24 seconds. If the train had eight coaches and two driving cars and so was 190 metres long, you might conclude that its speed was $190/4.24 = 44.8$ metres per second. However your value of 4.24 seconds for the time is likely to be seriously in error since it includes your reaction time. The fact that you can read your stop-watch to the nearest hundredth of a second does not mean that your timing is as accurate as that. If you were not very good at timing, it might be realistic to say that the time was between 4.0 and 4.4 seconds. The equivalent range of values for the train's speed is 47.5 to 43.2 metres per second. It would probably be safe to conclude that the speed of the train was between 43 and $47\frac{1}{2}$ metres per second.

Once you have given a mathematical explanation of your results, you must then check that your explanation will work with other data. If it does not, you may have to repeat your experiment, perhaps refined for greater accuracy, or to improve your explanation, or both.

For any real problem you should expect to go round one or both cycles in the flow chart at least twice and possibly several times before you home in on a solution that not only looks good on paper but actually matches the real world.

Give thought to the presentation of your solution. Pages of repetitive calculations are usually best summarised as a graph, data display or table, so that someone else can see your conclusions quickly and easily.

Do not include irrelevant information, but do not be afraid to show how you came upon your final method, including each cycle of the flow chart. State your conclusions clearly and in a place where they can easily be found.

In the following example the first approach is adopted, going straight into the mathematics of a simplified version of the problem.

EXAMPLE

A satellite 2000 km above the Earth is surveying the ground directly below using a camera with a 60° angle of view. How far apart can two objects be if they are to be in view at the same time?

Solution

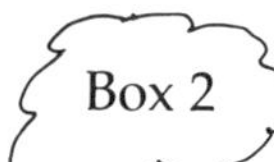

Assumptions

(a) Assume the two objects are on the ground.

(b) Assume the Earth is flat.

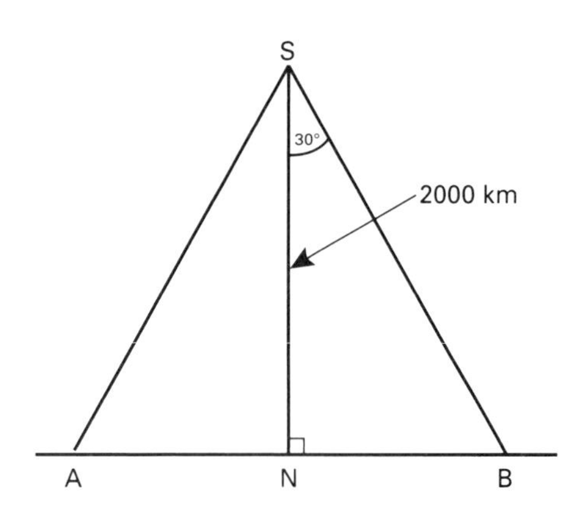

Boxes 3M and 4M

Mathematical representation See the diagram: S is the satellite, N the point on the ground directly below, A and B are the objects. A circular area is in view and AB is a diameter.

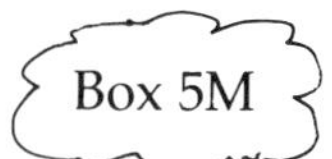

Manipulation

$$AB = 2NB = 2 \times 2000 \tan 30° \approx 2309 \text{ km}.$$

Check No direct check is possible. We can however test whether the result is reasonable by doing a scale drawing.

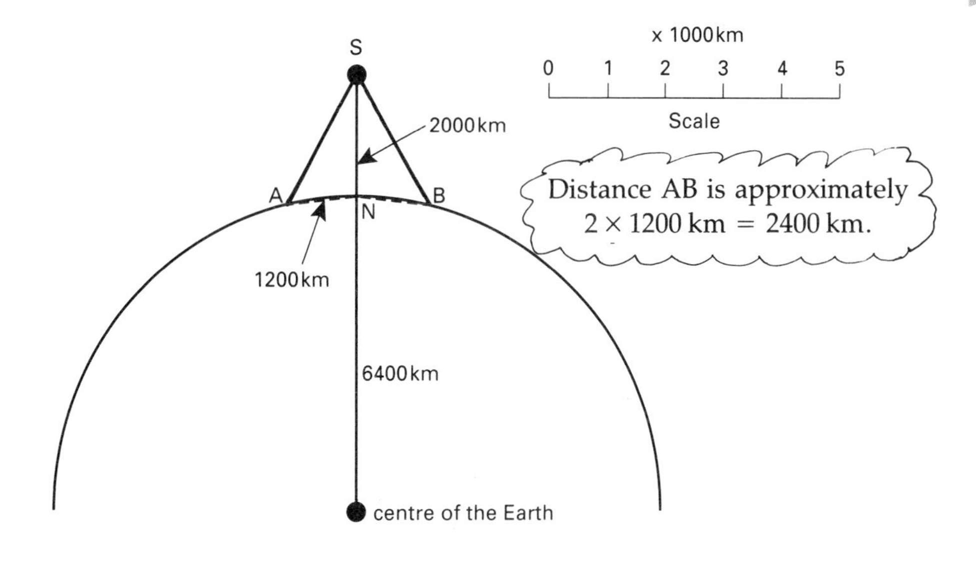

Box 6M

This shows the answer to be reasonable but also suggests a sensible way of refining the model.

Box 7M

The assumption 'The Earth is flat' is now replaced by 'The Earth is a perfect sphere, radius 6378 km'.

Box 8

Mathematical representation

A spherical cap is now in view: A and B are the ends of a diameter of the circular edge of the cap. The diagram shows a cross-section of the cap.

Box 3M

Manipulation

Using the sine rule on triangle SAC

$$\frac{6378}{\sin 30} = \frac{8378}{\sin \alpha}$$

$$\Rightarrow \sin \alpha = 0.6568 \text{ (to 4 significant figures)}$$

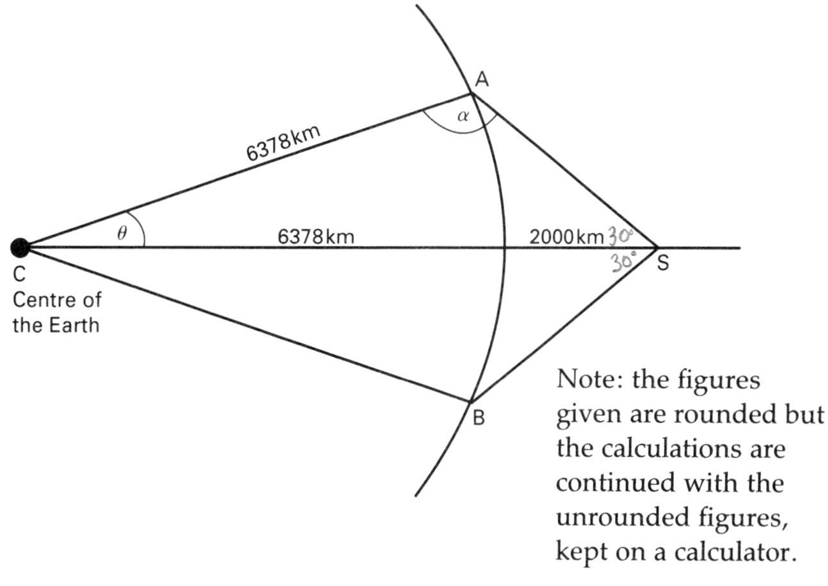

Note: the figures given are rounded but the calculations are continued with the unrounded figures, kept on a calculator.

$\alpha \approx 41°$ or $139°$ (triangle SAC is ambiguous).

However $\alpha = 41°$ gives $\theta > 90°$ which is not possible in this situation. Hence $\alpha = 139°$ and so $\theta = 11°$.

Arc length AB is $\dfrac{22}{360} \times 2 \times \pi \times 6378$ km ≈ 2461 km.

Check Again no direct check with reality is possible. However it is reassuring that the two answers (2309 km and 2461 km) are close together, and matched by the scale drawing estimate of 2400 km.

Answer We can be reasonably confident that two objects 2350 km apart can be viewed at the same time.

Further refinements The Earth is so nearly a perfect sphere that we are unlikely to want to refine this assumption.

The problem was posed with a satellite at 2000 km and a 60° angle of view. Most observation satellites are much higher than that (those in geostationary orbit are at 36 000 km) and use cameras with only a 1° or 2° field of view. A possible next step would be to generalise the result for a satellite at height h km and a camera with a field of view of $\theta°$.

Exercise 1A

Go through each of the three problems posed on p. 2 and discuss the assumptions you could make to model the situations. When making a simplifying assumption, try to assess the likely scale of its effect on the model.

You are **not** being asked to produce a mathematical model though you may, of course, do so if you wish.

2 Kinematics

Give me matter and motion and I will construct the Universe.

René Descartes, 1640

How would you describe the motion of the objects in the following photographs?

The aim of this chapter is to enable you to describe how objects move, including the way their movement is changing. We will not be concerned with why they are moving, but just describing how they move. This branch of mathematics is called *kinematics*.

When describing the motion of an object we consider three main features:

- the speed at which the object is moving, which could be either constant or changing;
- the direction in which it is moving, which could also be either constant or changing;
- whether every part of the object is moving in the same way.

In this chapter we concentrate on the simplest kind of motion: motion along a straight line.

Modelling real objects

Mechanics is concerned with the motion of objects and in many cases the motion is rather complicated. Take, for example, the motion of a leaf falling to the ground.

To describe its motion we need to do two things: first we must say where the leaf is, perhaps relative to its starting position, and secondly we must describe the orientation of the leaf. So the motion of the leaf could be described by specifying its overall motion along a path, which we call *translation* and its turning motion, which we call *rotation* or *spin*.

To start such a problem we simplify it, considering first the motion along a straight line. We ignore any rotations and follow the path of some *point* on or in the object. This is a very important simplification in mechanics. A point object is called a *particle*. It is a mathematical idealisation of an object in which all the matter of the object is assumed to occupy a single point in space. The particle has the mass of the object but no dimensions. Figure 2.1 shows two examples of the use of the particle model assumption.

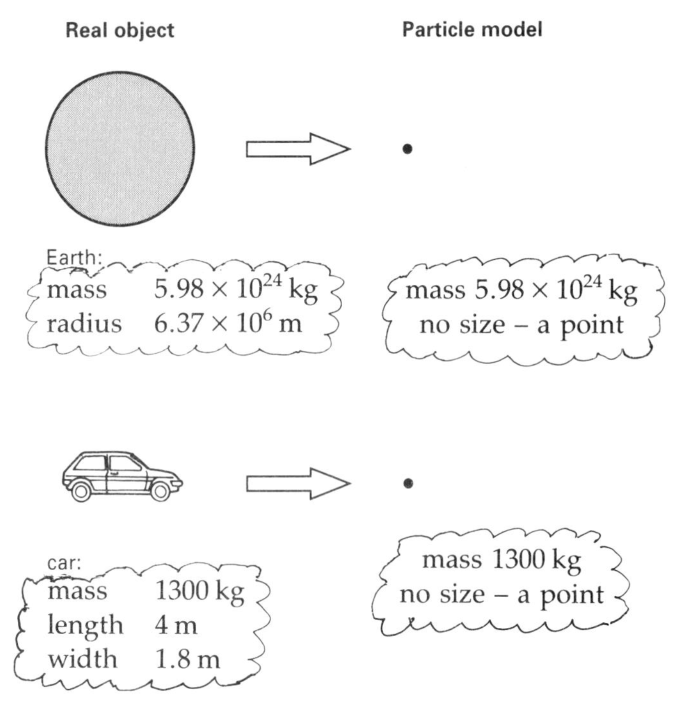

Figure 2.1 The particle model

You might ask whether there is a point in a real object which follows a path similar to that of the particle. In fact there is such a point and it is called the *centre of mass* of the object. So although we cannot use the particle model for a complete description of the motion of an object, we can use it to describe the motion of its centre of mass.

For Discussion

Which of the following could we model as the motion of a particle in a straight line? Where would you choose the 'point'?

(a) a table tennis ball in flight
(b) a train on its journey from Plymouth to London
(c) an aircraft coming in to land at Heathrow
(d) a ball rolling down a hill
(e) a snooker ball sliding over the table (a 'stun shot')
(f) a swimmer diving from a springboard.

Position, displacement and distance

When you are describing the motion of an object mathematically you usually start by defining an origin and axes. In one-dimensional motion the origin will be a point on the line along which the object is moving, and you will need to state which direction you are taking as positive. The *position* of the object is given relative to the origin, in a stated direction (figure 2.2).

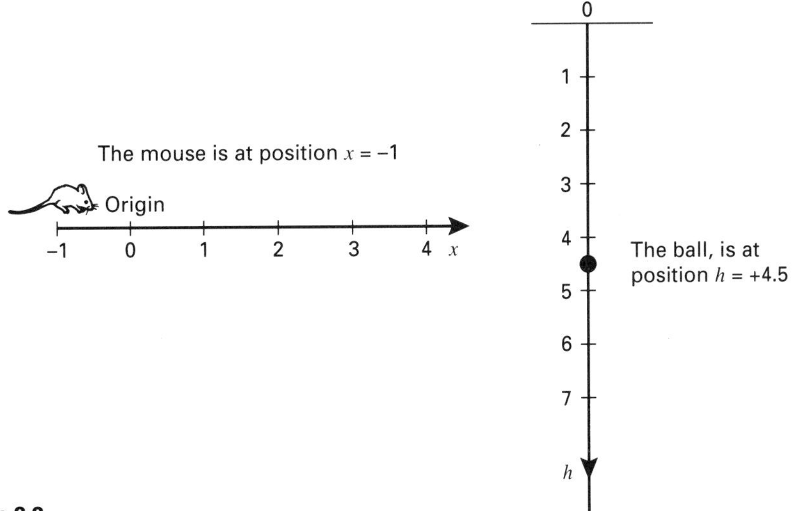

Figure 2.2

The change in the position of an object is called its *displacement*. Displacement is measured from a particular position in a stated direction. When displacement is measured from the origin it is the same as position.

The *distance* an object travels is the amount of ground it covers, irrespective of direction. In figure 2.3 the cat started at position +2 and ended at position +1. Its displacement over this period is −1 from its starting position +2. The distance it has travelled is 5 (from 2 to 4 and back to 1). Be careful not to confuse distance with displacement or position. All three terms are used in this book.

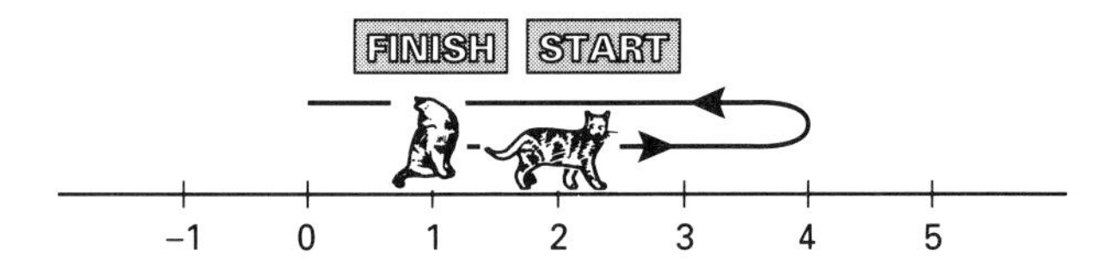

Figure 2.3

Speed and velocity

The word *speed* is one that you use often in everyday life. It describes the rate at which an object is covering distance. Like distance, it is given a value (like 90 mph or 40 ms^{-1}) but no direction.

The word *velocity* is used to describe speed in a stated direction. In one-dimensional motion the direction may be indicated by a plus or minus sign, or by some other description such as north or south.

In figure 2.4 the car has speed 80 mph and velocity +80 mph. The lorry has speed 30 mph and velocity −30 mph.

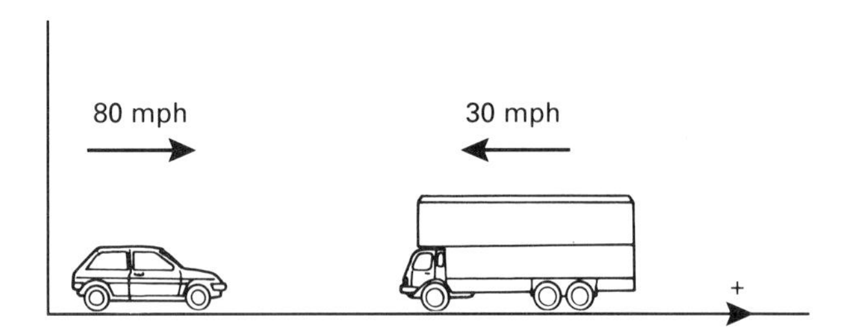

Figure 2.4

Notation and units

The letters s, h, x, y and z are all commonly used to denote position. The letter t is used to denote time measured from a starting instant. The position of a moving body depends on time and this is sometimes emphasised by writing s as $s(t)$, x as $x(t)$, etc.; when $t = 2$ the value of s would then be written $s(2)$. You read $s(t)$ as "the value of s at time t", and $s(2)$ as "the value of s at time $t = 2$".

In graphs of position, speed, velocity or acceleration against time, time always goes along the horizontal axis, as shown in figure 2.5.

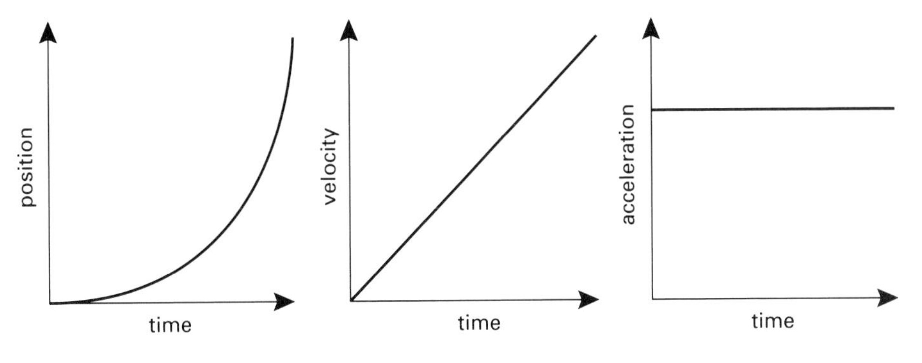

Figure 2.5

The S.I. (Système International d'Unités) unit for distance is the metre (m), that for time is the second (s) and that for mass the kilogram (kg). Other units follow from these so speed is measured in metres per second, written ms^{-1}

S.I. units are used almost entirely in this book, but occasional references are made to Imperial and other units.

Exercise 2A

1. What are the positions of the particles A, B and C in the figure below?

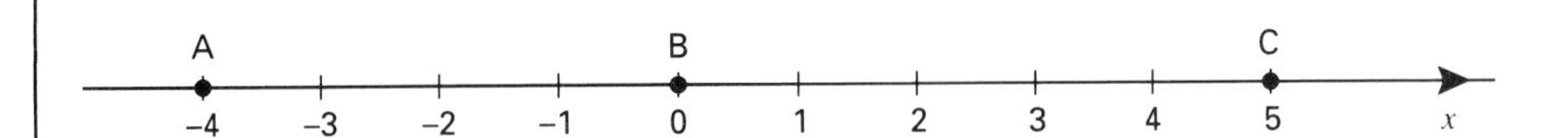

What is the displacement of B (i) relative to A, (ii) relative to C?

2. The position of a particle is given by

$$x = 2 + t(t - 3) \qquad (0 \leqslant t \leqslant 5)$$

where x is measured in metres and t in seconds.

(i) What is the position of the particle at times $t = 0$, 1, 2 and 3?

(ii) Draw a diagram showing the position of the particle at these times.

(iii) Find the displacement of the particle relative to its initial position at $t = 5$.

(iv) Calculate the total distance travelled during the motion.

3. For each of the following situations sketch a graph of position against time.

(i) A stone dropped from a bridge which is 40 metres above a river.

(ii) A parachutist who jumps from a helicopter hovering at 2000 metres and opens his parachute after 10 seconds of 'free fall'.

(iii) A 'bungee jumper' on the end of an elastic string.

Exercise 2A continued

4. A boy throws a stone vertically upwards so that its position y m at time t s is as shown in the figure.

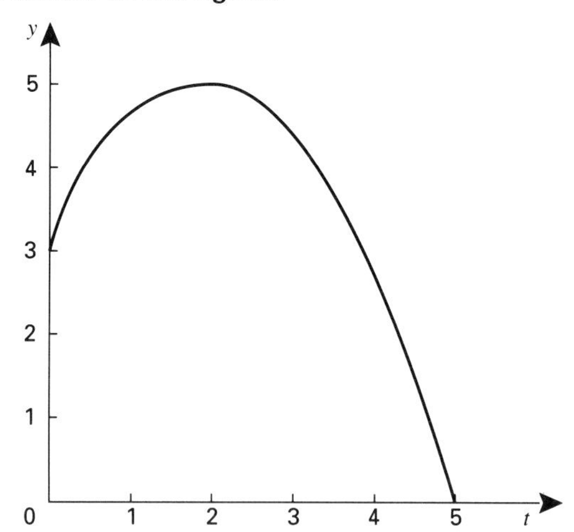

(i) From the graph write down the position of the ball at times $t = 0, 1, 2, 3, 4, 5$.

(ii) Calculate the displacement of the ball relative to its starting position at times $t = 1, 2, 3, 4, 5$.

(iii) What is the total distance travelled
  (a) during the first 2 seconds,
  (b) during the 5 seconds of the motion?

5. A particle moves so that its position in metres at time t seconds is

$$x = 2t^3 - 18t$$

(i) Calculate the position of the particle at times $t = 1, 2, 3$ and 4.

(ii) Sketch a graph of the position against time.

(iii) Find the times when the particle is at the origin and describe in which direction the particle is moving.

6. A stone is thrown upwards from a window so that its height above the ground is modelled by

$$h = 4 + 0.9t - 4.9t^2$$

(i) When does the stone hit the ground?

(ii) What is the displacement of the stone relative to the window when it hits the ground?

(iii) By drawing a graph of h against t, estimate the total distance travelled by the stone.

Average Speed

In practice speeds often vary continuously. For example, on a car journey it is almost impossible to drive at a constant speed for any length of time. To describe the speed of an object such as a car would be very complicated. The idea of average speed can be used to give a simple description of the speed of an object over a period of time. It is defined as

$$\text{average speed} = \frac{\text{total distance travelled}}{\text{time taken}}.$$

In fact average speed is the speed at which an object would have travelled if it had completed its journey in the same time, but at constant speed.

EXAMPLE A car travels 100 miles in 2½ hours. Find the average speed of the car.

Solution

$$\begin{aligned}\text{average speed} &= \frac{\text{total distance travelled}}{\text{time taken}}\\ &= \frac{100}{2.5}\\ &= 40 \text{ mph.}\end{aligned}$$

EXAMPLE A train travels at a constant 120 kmh^{-1} for 30 minutes and then covers the next 100 km in 40 minutes. Find the average speed for the whole journey.

Solution

For the first part of the journey the average speed is 120 kmh^{-1} and so the train covers 60 km in 30 minutes.

For the whole journey the average speed is given by:

$$\begin{aligned}\text{average speed} &= \frac{\text{total distance travelled}}{\text{time taken}}\\ &= \frac{60 + 100}{30 + 40}\\ &= 2.29 \text{ km per minute}\\ &= 137 \text{ kmh}^{-1}\end{aligned}$$

Exercise 2B

1. Nigel Mansell won the British Grand Prix in 1992 by completing the 308 km in 1 hour 25 minutes 43 seconds.
   (i) Calculate his average speed in ms^{-1}.
   (ii) His nearest rival finished 40 seconds later. Calculate his average speed.

Exercise 2B continued

2. A train on a Cornish branch line takes 45 minutes to complete its 15 mile trip. It stops for 3 minutes at each of 7 stations during the trip.
   (i) Calculate the average speed of the train.
   (ii) What increase in average speed could be achieved by reducing the stop at each station to 2 minutes?

3. An aeroplane flies from London to Toronto, a distance of 3560 miles at an average speed of 800 mph. It returns at an average speed of 750 mph. Find the average speed for the round trip.

4. In a medley swimming relay, teams of 3 take part, each member swimming 100 m. The first member of the team has an average speed of 1.8 ms^{-1}. The second takes 10 s longer than the first and the last swimmer takes 50 seconds. Find the average speed for the whole swim.

5. A fun runner enters a 5 km race. During the first 2 km the runner has an average speed of 3.5 ms^{-1}, during the next 2 km an average speed of 2.8 ms^{-1} and during the final kilometre an average speed of 3 ms^{-1}. Find the average speed for the whole run and the time taken to complete the race.

Graphs

It is often helpful to draw graphs to illustrate motion.

Distance–time and speed–time graphs

A car is travelling along a road at a constant speed of 20 ms^{-1}. The distance–time and speed–time graphs are shown in figure 2.6. The speed of the car is given by the gradient of the line in the distance–time graph.

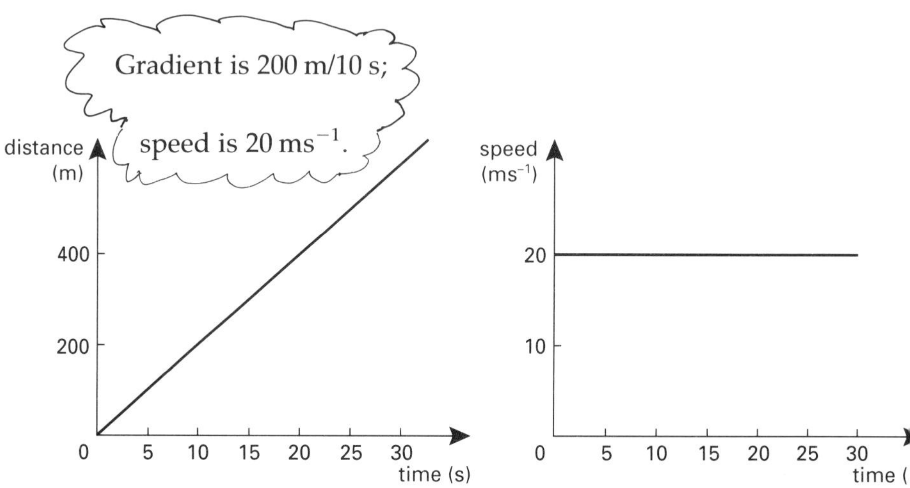

Figure 2.6

A car is unlikely to be travelling in an absolutely straight line since all roads have curves in them, so distance and speed are more appropriate measures in this case than position and velocity.

Position–time and velocity–time graphs

The position–time graph in figure 2.7 shows the position of a ball t seconds after being thrown vertically upwards with velocity 20 ms^{-1}. You can find the gradient at any point, as is done here at (1,15). In this case the gradient is found to be 10 m $\div$ 1 s corresponding to a velocity of $+10\ ms^{-1}$.

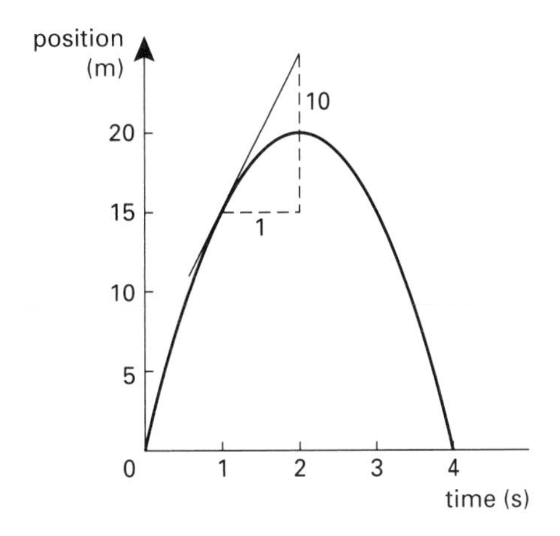

Figure 2.7

Finding the gradient at several points allows you to plot the velocity–time graph. In this case it is a straight line (figure 2.8). Notice that the speed–time graph for the same situation is different since speed can never be negative. This graph (figure 2.9) gives you less information.

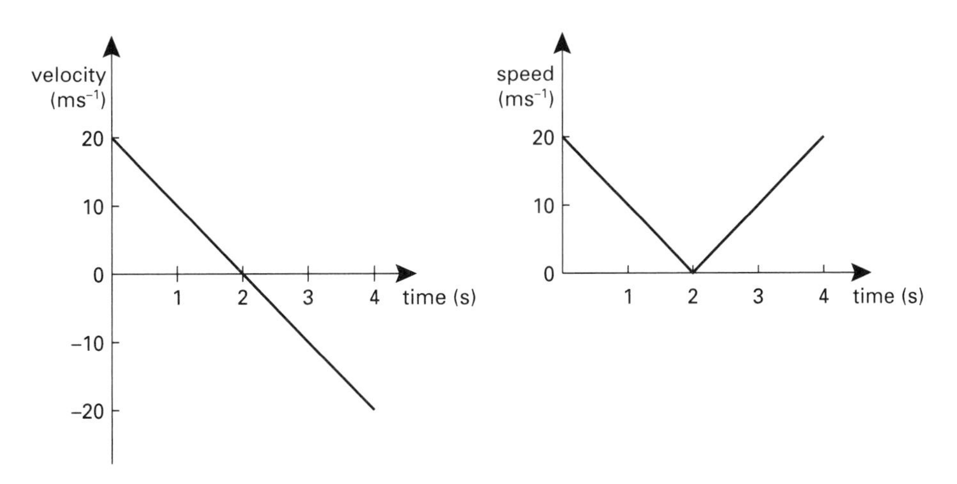

Figure 2.8

Figure 2.9

The relationship between velocity and position

In some situations you will know the position, s, of an object at any time, t, in the form of an algebraic expression in terms of t. When this is the case you can also find the gradient of the position–time graph (and so the object's velocity) by differentiating:

$$\text{velocity, } v = \frac{\mathrm{d}s}{\mathrm{d}t}.$$

EXAMPLE

The distance travelled by a car from a set of traffic lights is modelled by

$$s = 1.5t^2 + 3t$$

where t is time in seconds and s is distance in metres.

(i) Draw a distance–time graph of the motion for $t \leqslant 5$.
(ii) Find the speed of the car as a function of t.
(iii) Draw a speed–time graph of the motion for $t \leqslant 5$.
(iv) Describe the motion of the car.

Solution

(i)

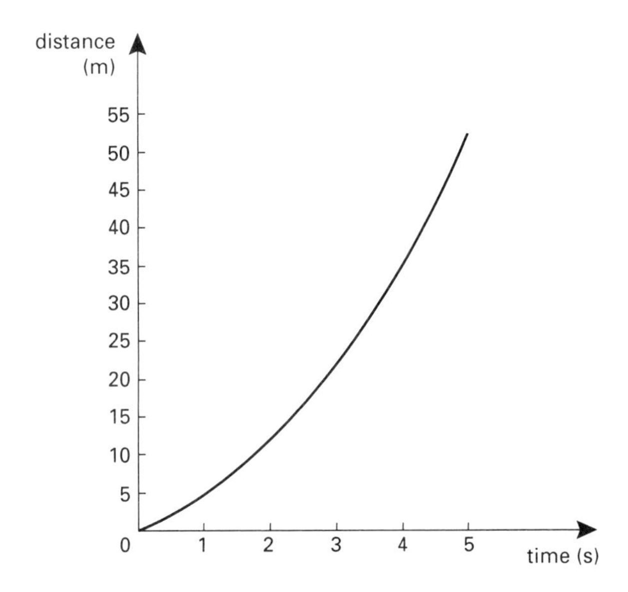

(ii) The speed of the car is given by $\frac{\mathrm{d}s}{\mathrm{d}t}$:

$$v = \frac{\mathrm{d}s}{\mathrm{d}t} = 3t + 3.$$

(iii) The speed–time graph is the straight line $v = 3t + 3$.

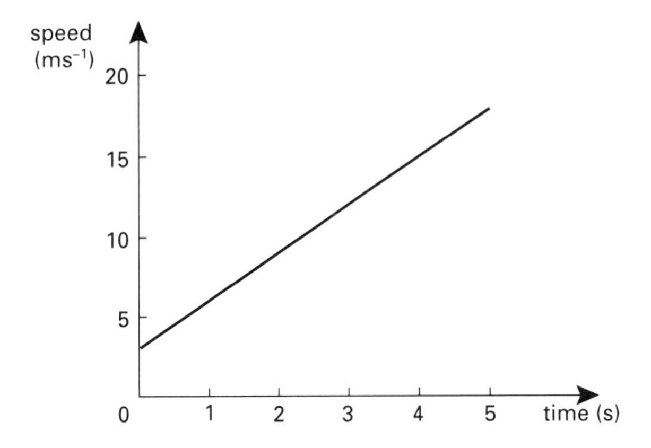

(iv) The car begins with speed 3 ms^{-1} and in 5 s travels 52.5 m. Its speed increases at a constant rate to 18 ms^{-1} during the 5 second interval.

Exercise 2C

1. The distance–time graph shows the relationship between the distance travelled and time for a person who leaves home at 9.00 am, walks to the bus stop and catches a bus into town.

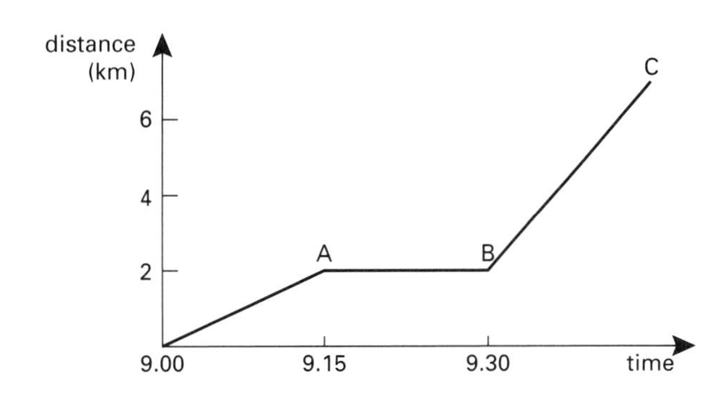

(i) Describe what is happening during the time from A to B.
(ii) The section BC is much steeper than OA; what does this tell us about the motion?
(iii) Draw the speed–time graph for the person.
(iv) What simplifications have been made in drawing these graphs?

2. A car is being driven along a road. It travels initially at 30 mph but over a 10 second period slows down at a constant rate to 5 mph to cross a 'sleeping policeman'.

Draw distance–time and speed–time graphs for the car.

Exercise 2C continued

3. The diagram shows a velocity–time graph. Sketch the corresponding position–time graph, given that the object starts at the origin.

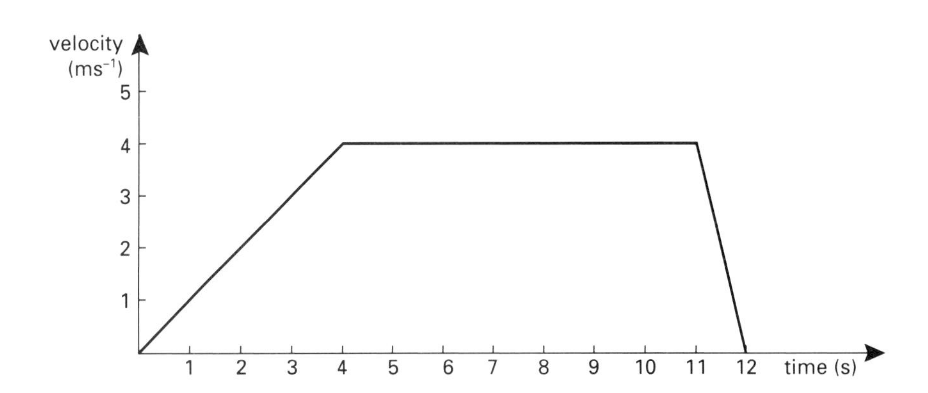

4. A cyclist starts from rest and increases speed uniformly for 10 s, then travels at a constant speed for 30 s, before slowing down and stopping in 15 s. Draw distance–time and speed–time graphs for the cyclist.

5. As a lorry goes up a steep hill, its speed decreases to a minimum and then remains constant. Sketch speed–time and distance–time graphs for the lorry.

6. Long distance runners use a training technique known as 'scouts' pace' to improve their speed. This involves alternating between sprinting and running for equal distances with no rests in between. Sketch distance–time and speed–time graphs for a runner training in this way.

7. The figure shows a distance–time graph for two cars A and B.

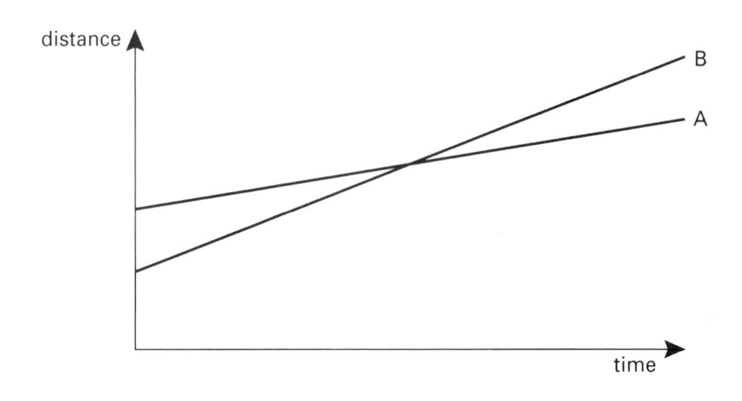

(i) Does car A ever move faster than car B?
(ii) Do the cars ever have the same speed?

Exercise 2C continued

8. A stone is dropped from a window 5 m above ground level. The distance in metres travelled by the stone in t seconds is modelled by

$$s = 5t^2$$

(i) How long does it take the stone to reach the ground?
(ii) Find the speed of the stone in terms of t.
(iii) What is the speed of the stone when it hits the ground?

9. The distance in metres travelled by a car as it moves off from rest is modelled by

$$s = \frac{t^3}{10} + \frac{t^2}{5} + \frac{t}{2}$$

where t is in seconds, for the first 10 seconds.
(i) How far does the car move during the first 10 seconds?
(ii) What is its speed after 10 seconds?
(iii) When is the speed 20 ms^{-1}?

10. A jig-saw blade moves so that the displacement in cm of the tip of the blade from the base of the saw guide at time t seconds is modelled by

$$s = 10^6 \times t(2.5 \times 10^{-3} - t) + 4$$

for $0 \leqslant t \leqslant 2.5 \times 10^{-3}$.
Find the maximum speed of the saw blade within this time interval.

Experiment

Tilt the surface of a table a little so that a small ball, such as a marble or ball bearing, will roll slowly down it. On a sheet of paper mark approximately 10 parallel lines 10 cm apart and fix the paper to the inclined table.

Release the ball at the top mark and make a note of the time as the ball passes each line.

- Draw position–time and velocity–time graphs.
- Find a relationship between the distance travelled, s, and the time t.

Finding distance from speed

For Discussion

These distance–time and speed–time graphs model the motion of a stone falling from rest for 3 seconds.

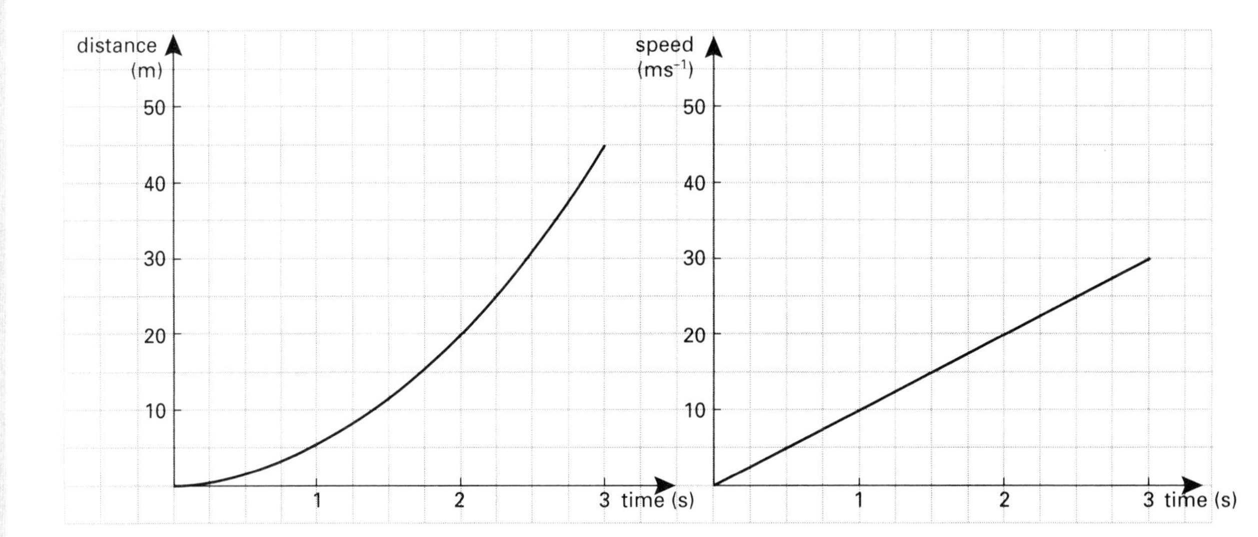

Calculate the area under the speed–time graph between (i) $t = 0$ and 1, (ii) $t = 0$ and 2, (iii) $t = 0$ and 3.

Compare your answers with the distance that the stone has fallen, shown on the distance–time graph at $t = 1$, 2 and 3.

What conclusions do you reach?

The area under a speed–time graph represents the distance travelled.

An equivalent result holds for velocity–time graphs.

The area under a velocity–time graph represents the change in position, or the displacement.

The areas under the speed–time and velocity–time graphs are the same if the velocity is in the positive direction but different if it is negative. In that case the area will be below the axis and so represent a negative displacement.

EXAMPLE A man walks east for 6 seconds at 2 ms^{-1} then west for 2 seconds at 1 ms^{-1}. Draw (i) the speed–time graph and (ii) the velocity–time graph, and interpret the area underneath each.

Solution

The man's journey is illustrated below.

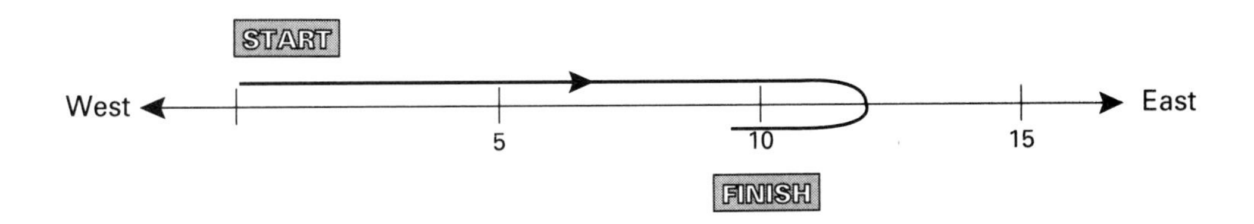

(i) Speed–time

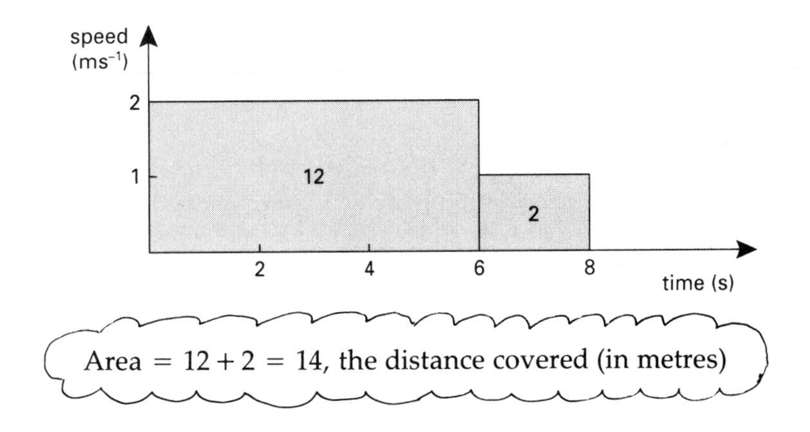

Area $= 12 + 2 = 14$, the distance covered (in metres)

(ii) Velocity–time

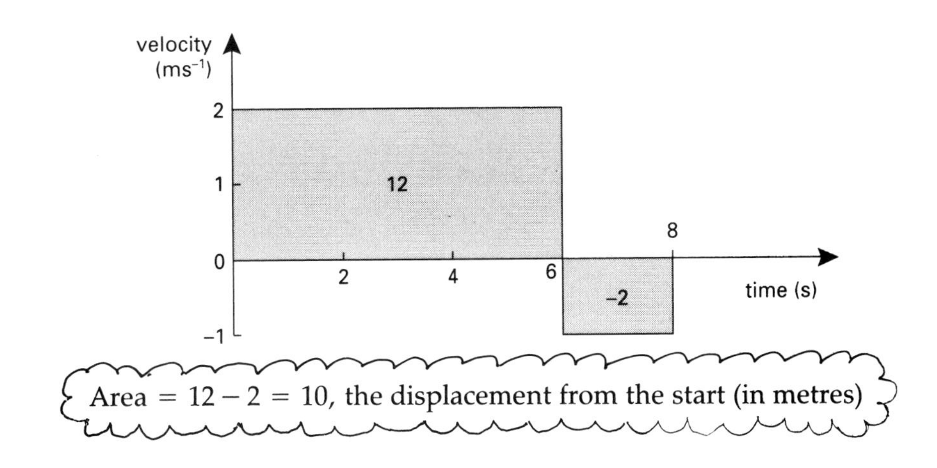

Area $= 12 - 2 = 10$, the displacement from the start (in metres)

Finding the area under velocity–time and speed–time graphs

(1) Triangles, rectangles and trapezia

Many of these graphs consist of straight line sections and so the area is easily found by splitting it up into triangles, rectangles or trapezia (figure 2.10).

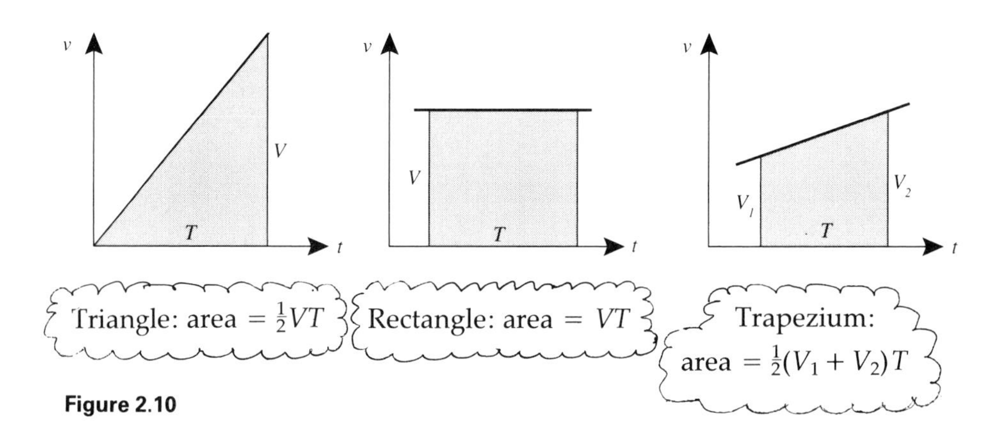

Figure 2.10

EXAMPLE

Robin is cycling home. He turns off the main road at 4 ms^{-1} and accelerates uniformly to 10 ms^{-1} over the next 6 seconds. He maintains this speed for 20 seconds and then slows down uniformly for 4 seconds to stop outside his house.

(i) Draw the speed–time graph for his journey.

(ii) How far does he live from the main road?

Solution

(i)

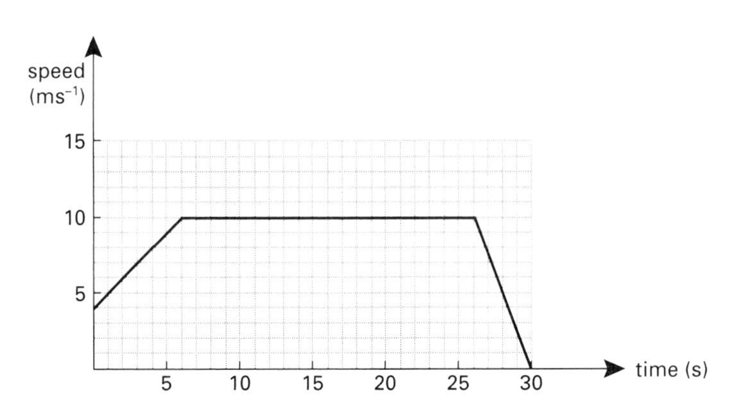

(ii) The area under the speed–time graph is found by splitting it into three regions.

A trapezium: area $= \frac{1}{2}(4 + 10) \times 6 = 42$ m

B rectangle: area $= 10 \times 20 = 200$ m

C triangle: area $= \frac{1}{2} \times 10 \times 4 = 20$ m

total area $= 262$ m

Robin lives 262 m from the main road.

---

(2) Integration

If you know an algebraic expression for the velocity, v, at time t, you can find the area under a velocity–time graph by integration (see figure 2.11).

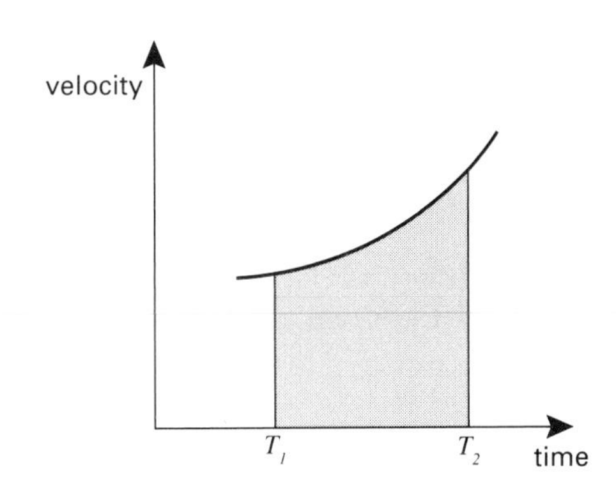

Figure 2.11

$$\text{Area} = \int_{T_1}^{T_2} v\,\mathrm{d}t$$

EXAMPLE

A car moves between two sets of traffic lights, stopping at both, with speed v ms^{-1} at time t s modelled by

$$v = \frac{1}{20}t(40 - t) \qquad 0 \leqslant t \leqslant 40$$

Find the times at which the car is stationary and the distance between the two sets of traffic lights.

Solution

The car is stationary when $v = 0$. Substituting this into the expression for the speed gives:

$$0 = \frac{1}{20}t(40 - t)$$

The solutions of this equation are $t = 0$ and $t = 40$. These are the times when the car starts to move away from the first set of traffic lights and stops at the second set. Hence the distance between the two sets of lights is given by:

$$\text{Distance} = \int_0^{40} \frac{1}{20} t(40 - t)\, dt$$

$$= \frac{1}{20}\left[20t^2 - \frac{t^3}{3}\right]_0^{40}$$

$$= 533\tfrac{1}{3}\text{ m}$$

(3) Estimation

Sometimes the velocity–time graph does not consist of straight lines, nor do you know an algebraic expression for it. In that case you have to make the best estimate you can, by counting the squares underneath it, or by replacing the curve by a number of straight lines.

For Discussion

This speed–time graph shows the motion of a dog over a 60 second period. Estimate how far the dog travelled during this time.

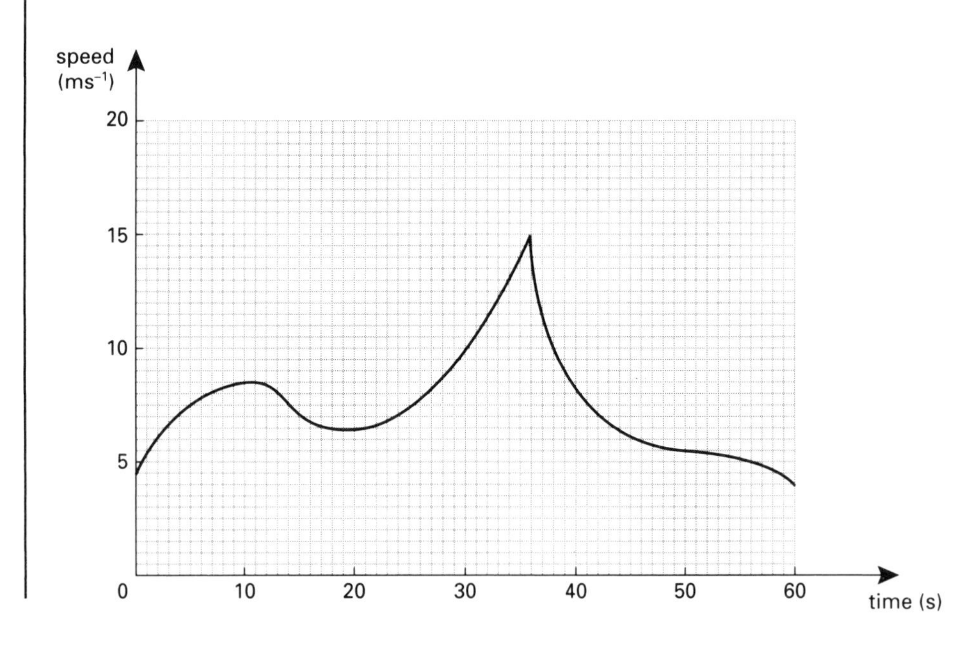

Exercise 2D

1. The speed of a ball rolling down a hill is modelled by

$$v = 1.7t \qquad (\text{in ms}^{-1}).$$

(i) Draw the speed–time graph of the ball.

(ii) How far does the ball travel in 10 seconds?

2. Ted, John and Sharon have initial positions $S_T(0) = 0$, $S_J(0) = 52$ and $S_S(0) = 100$ respectively. Ted runs at 7 ms^{-1}, John jogs at 5 ms^{-1} and Sharon walks at 2 ms^{-1}.

(i) Who catches whom first?

(ii) The position of the bar in the Dog and Whistle pub is $S_B = 210$. The first to reach the bar orders the drinks while the last to arrive pays. Who orders, and who pays?

3. A car is moving at 20 ms^{-1} when it begins to increase speed, so that every 60 seconds it gains 5 ms^{-1} until it reaches its maximum speed of 50 ms^{-1}.

(i) Draw the speed–time graph of the car.

(ii) Write down an expression for the speed of the car t seconds after it begins to speed up.

(iii) When does the car reach its maximum speed of 50 ms^{-1}?

(iv) Find the distance travelled by the car after 150 s.

4. A train leaves a station where it has been at rest and picks up speed at a constant rate for 60 seconds. It then remains at a constant speed of 17 ms^{-1} for 60 s before it begins to slow down uniformly to 10 ms^{-1} as it approaches a set of signals. After 45 s the signal changes and the train again increases speed uniformly for 75 s until it reaches a speed of 20 ms^{-1}. A second set of signals then orders the train to stop, which it does after slowing down uniformly for 30 s. Draw a speed–time graph for the train and use this to find the distance that it has travelled from the station.

5. When a parachutist jumps from a helicopter hovering above an airfield her speed increases at a constant rate to 28 ms^{-1} in the first 3 seconds of her fall. It then decreases uniformly to 8 ms^{-1} in a further 6 seconds, remaining constant until she reaches the ground.

(i) Sketch a speed–time graph for the parachutist.

(ii) Find the height of the aeroplane when the parachutist jumps out if the complete jump takes 1 minute.

6. The speed of a bullet t seconds after entering water is modelled by

$$v = 216 - t^3 \quad (\text{in ms}^{-1})$$

until it stops moving.

(i) When does the bullet stop moving?

(ii) How far has it travelled by this time?

7. During braking the speed of a car is modelled by

$$v = 40 - 2t^2 \quad (\text{in ms}^{-1})$$

until it stops moving.

(i) How long does the car take to stop?

(ii) How far does it move before it stops?

Exercise 2D continued

8. A boy throws a ball up in the air from a height of 1.5 m and catches it at the same height. Its height in m at time t s is

$$s = 1.5 + 15t - 5t^2$$

(i) What is the vertical velocity v of the ball at time t?
(ii) Find the position, velocity and speed of the ball at $t = 1$ and $t = 2$.
(iii) Sketch the position–time, velocity–time and speed–time graphs for $0 \leqslant t \leqslant 3$.
(iv) When does the boy catch the ball?
(v) Explain why the distance travelled by the ball is not equal to $\int_0^3 v\,dt$, and state what information this expression does give.

9. An object moves along a straight line so that its position in metres at time t seconds is given by

$$s = t^3 - 3t^2 - t + 3 \qquad (t \geqslant 0).$$

(i) Find the position, velocity and speed of the object at $t = 2$.
(ii) Find the smallest time when (a) the position is zero, and (b) the velocity is zero.
(iii) Sketch position–time, velocity–time and speed–time graphs for $0 \leqslant t \leqslant 4$.
(iv) Describe the motion of the object.

10. An object moves along a straight line so that its velocity at time t seconds is modelled by

$$v = 16 - 8t \text{ (in ms}^{-1}) \qquad (t \geqslant 0).$$

It starts its motion at the origin.
(i) Find a formula for the position of the object at time t.
(ii) Find the smallest non-zero time when (a) the velocity is zero, and (b) the object is at the origin.
(iii) Sketch the position–time, velocity–time and speed–time graphs for $0 \leqslant t \leqslant 4$.

11. A stone is thrown upwards from a window, so that t seconds later its height (in metres) above the ground is modelled by

$$s = 4 + 2t - 4.9t^2.$$

(i) When does the stone hit the ground?
(ii) Find an expression for the velocity of the ball.
(iii) How fast is the stone moving when it hits the ground?

12. Two objects move along the same straight line. The velocities of the objects (in ms^{-1}) are given by $v_1 = 16t - 6t^2$ and $v_2 = 2t - 10$ $(t \geqslant 0)$. Initially the objects are 32 metres apart. At what time do the objects collide?

Acceleration

We have already used the term velocity to describe how the position of an object is changing. The term *acceleration* is used to describe how the *velocity* of an object is changing.

In everyday language the term acceleration is often associated only with increasing speed, but in mechanics it is used to describe rate of change of velocity. It can therefore be applied when the magnitude of the velocity is decreasing rather than increasing, or when the direction of the velocity is changing.

In this chapter we are looking at motion in one dimension only, so whenever you meet acceleration it will be due to an increase or decrease in speed. In Chapter 4, where motion in two and three dimensions is considered, acceleration may be due to a change in direction, as in the case of a car being driven at constant speed round a roundabout.

The S.I. unit for acceleration is the metre per second per second, written ms^{-2}.

A familiar example is that of a falling object accelerating due to gravitational attraction. You know that if you drop a stone its speed increases; the rate at which it increases is 9.81 ms^{-2}, a value often approximated by 9.8 or 10 ms^{-2}. Gravitational acceleration is denoted by g.

If a velocity is constant there is no acceleration, since the velocity is not changing, but in all situations where the velocity is changing there is an acceleration.

For Discussion

Are the following objects accelerating? If you think they are accelerating, explain why you have come to this conclusion.

(i) A lorry pulling away from a set of traffic lights on a straight road.
(ii) A cyclist travelling along a winding lane at a constant speed.
(iii) An aeroplane ascending at constant speed along a straight line.
(iv) A parachutist from the moment of jumping to the time of landing.
(v) A planet orbiting the Sun.
(vi) A bungee jumper from the moment of jumping until he finally stops moving.

Acceleration–time graphs

When looking at speed we started with the distance–time graph and from the slope of the graph at various points formed the speed–time graph. Alternatively we could use calculus:

$$v = \frac{ds}{dt}.$$

For acceleration we follow a similar route. The rate of change of speed or the gradient of the speed–time graph at a particular instant gives the acceleration at that instant.

For one dimensional motion, acceleration is given by

$$a = \frac{dv}{dt} = \frac{d^2s}{dt^2}$$

If the acceleration is a known function of time then the velocity can be found by using integration:

$$v = \int a \, dt.$$

EXAMPLE

An object moves along a straight line so that its position at time t in seconds is given by

$$s = 2t^3 - 6t \text{ (in metres)} \qquad (t \geqslant 0).$$

(i) Find the acceleration of the object at $t = 0$, 1 and 2.
(ii) Sketch the acceleration–time graph.
(iii) Describe the motion of the object.

Solution

Position $\qquad s = 2t^3 - 6t$

Velocity $\qquad v = \frac{ds}{dt} = 6t^2 - 6$

Acceleration $\qquad a = \frac{dv}{dt} = 12t$

(i) At $t = 0$, 1 and 2 we have $a(0) = 0$, $a(1) = 12$ and $a(2) = 24$.
(ii)

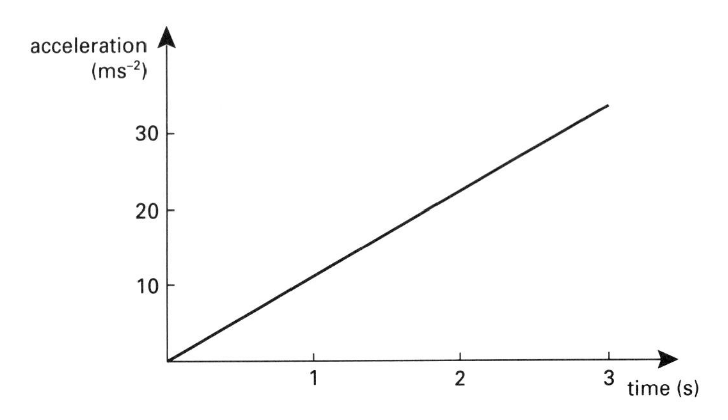

(iii) The acceleration is increasing in magnitude and is always in the direction of s increasing. The initial velocity is negative so the object begins at the origin and moves to the left with initial speed 6 ms^{-1}. It slows down to stop at $s = -4$. It then moves to the right with increasing speed.

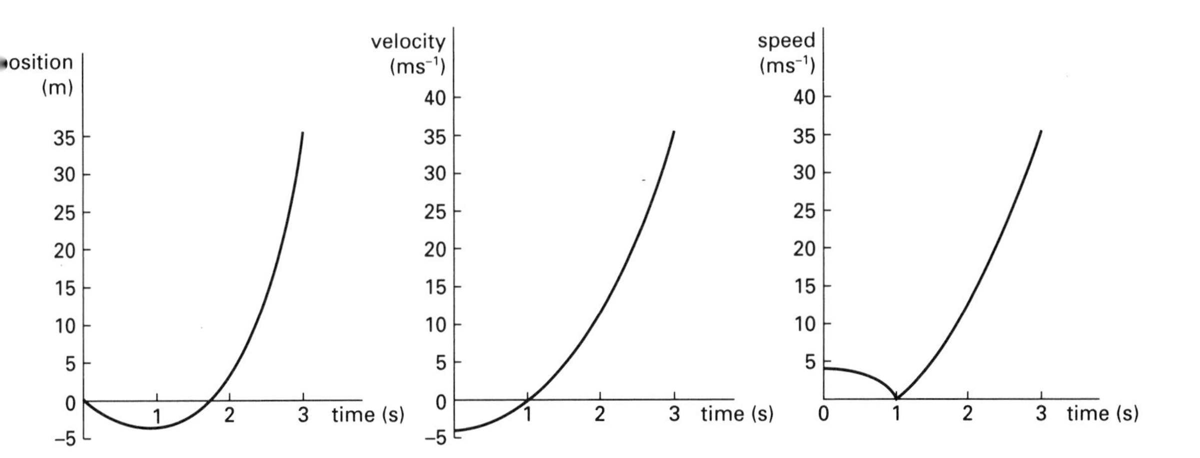

EXAMPLE

The acceleration of a particle (in ms^{-2}) at time t seconds is given by

$$a = 6 - t.$$

The particle is initially at the origin, with velocity $-2\ ms^{-1}$.
Find (i) the speed of the particle 6 seconds later;
(ii) the position of the particle 6 seconds later.

Solution

The information given may be summarised as follows:

$$\text{at } t = 0,\ v = -2 \text{ and } s = 0;\ a = \frac{dv}{dt} = 6 - t$$

(i) Integrating gives $\quad v = 6t - \tfrac{1}{2}t^2 + c$

Substituting $v = -2$, $t = 0$ gives $-2 = 0 - 0 + c \Rightarrow c = -2$

So $\quad v = -2 + 6t - \tfrac{1}{2}t^2.$

When $t = 6$ $\quad v = -2 + 36 - 18 = 16$

The speed of the particle after 6 seconds is $+16\ ms^{-1}$.

(ii) $$v = \frac{ds}{dt} = -2 + 6t - \tfrac{1}{2}t^2$$

Integrating gives $s = -2t + 3t^2 - \tfrac{1}{6}t^3 + k$

Substituting $s = 0,\ t = 0$ gives $0 = -0 + 0 - 0 + k \quad \Rightarrow k = 0$

So $s = -2t + 3t^2 - \tfrac{1}{6}t^3.$

When $t = 6$ $\quad s = -12 + 108 - 36 = 60$

After 6 seconds the particle's position is +60 m.

For Discussion

We have defined velocity in terms of position by $v = \frac{ds}{dt}$ and acceleration in terms of velocity by $a = \frac{dv}{dt}$. What do you know about the motion of an object for which $\frac{ds}{dt} < 0$ and/or $\frac{dv}{dt} < 0$?

Exercise 2E

1. For each case below find the acceleration at $t = 1$ and $t = 2$. Draw the acceleration–time graph and describe how the velocity and acceleration change during the motion of each object.
(i) $x = 15t - 5t^2$
(ii) $x = 6t^3 - 18t^2 - 6t + 3$

2. The distance travelled by a cyclist is modelled by

$$s = 4t + 0.5t^2 \text{ in S.I. units.}$$

Find the acceleration of the cyclist at time t.

3. An object moves along a straight line so that its acceleration in ms^{-2} is given by $a = 4 - 2t$. It starts its motion at the origin with speed 4 ms^{-1} in the direction of x increasing.
(i) Find as functions of t the position and velocity of the object.
(ii) Sketch the position–time, velocity–time and acceleration–time graphs for $0 \leqslant t \leqslant 2$.
(iii) Describe the motion of the object.

4. A car accelerates away from a set of traffic lights. It accelerates to a maximum speed and at that instant starts to slow down to stop at a second set of lights. Which of the graphs below could represent (i) a distance–time graph, (ii) a velocity–time graph, (iii) an acceleration–time graph of its motion?

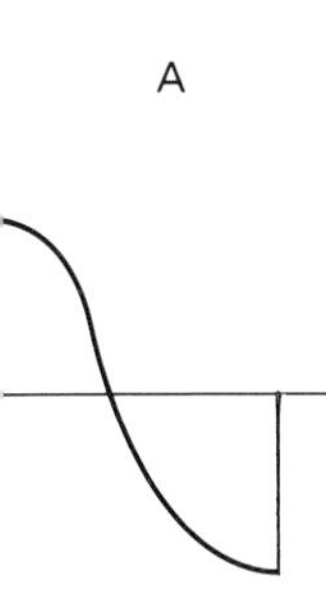

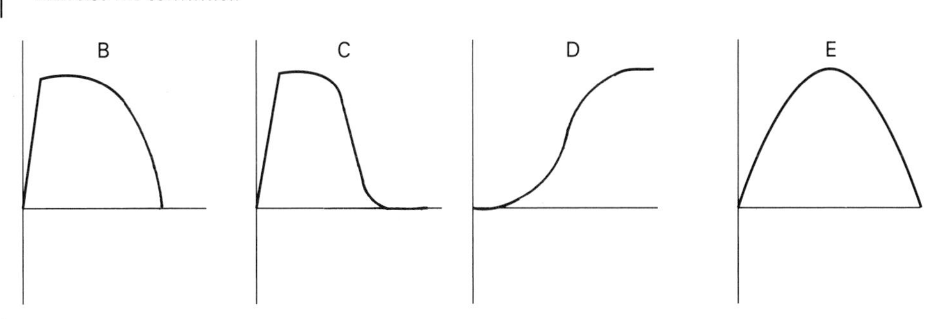

5. A film of a dragster doing a 400 m run from a standing start yields the following positions at 1 second intervals.

   Draw (i) a displacement–time graph, (ii) a velocity–time graph, (iii) an acceleration–time graph of its motion.

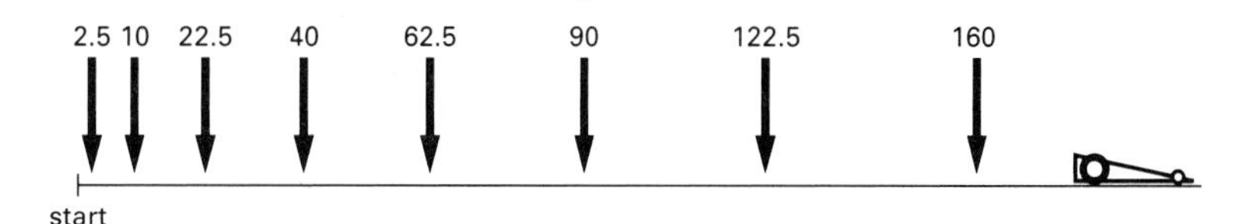

6. In each case below, the object moves along a straight line with acceleration a in ms^{-2}. Find an expression for the velocity v and position x of each object at time t s.
   (i) $a = 10 + 3t - t^2$: the object is initially at the origin and at rest.
   (ii) $a = 4t - 2t^2$: at $t = 0$, $x = 1$ and $v = 2$.
   (iii) $a = 10 - 2t$: at $t = 0$, $x = 0$ and $v = -5$.

7. Nick watches a golfer putting her ball 24 metres from the edge of the green and into the hole, and he decides to model the motion of the ball. Assuming that the ball is a particle travelling along a straight line, he models its distance, s metres, from the golfer at time t seconds by

$$s = -\frac{3}{2}t^2 + 12t \qquad 0 \leqslant t \leqslant 4.$$

   (i) Find the value of s when $t = 0, 1, 2, 3$ and 4.
   (ii) Explain the restriction $0 \leqslant t \leqslant 4$.
   (iii) Find the velocity of the ball at time t seconds.
   (iv) With what speed does the ball enter the hole?
   (v) Find the acceleration of the ball at time t seconds.

8. Andrew and Elizabeth are having a race over 100 metres. Their accelerations are as follows, in ms^{-2}:

Andrew		**Elizabeth**	
$a = 4 - 0.8t$	$0 \leqslant t \leqslant 5$	$a = 4$	$0 \leqslant t < 2.4$
$a = 0$	$t > 5$	$a = 0$	$t > 2.4$

   (i) Find the greatest speed of each runner.
   (ii) Sketch the speed–time graph for each runner.

(iii) Find the distance Elizabeth runs while reaching her greatest speed.

(iv) How long does Elizabeth take to complete the race?

(v) Who wins the race, by what time margin and by what distance?

On another day they race over 120 metres, both running in exactly the same manner.

(vi) What is the result now?

9. Christine is a parachutist. On one of her descents, her vertical speed, v ms^{-1}, t seconds after leaving an aircraft is modelled by

$v = 8.5t$ $\quad 0 \leqslant t \leqslant 10$

$v = 5 + 0.8(t - 20)^2$ $\quad 10 < t \leqslant 20$

$v = 5$ $\quad 20 < t \leqslant 90$

$v = 0$ $\quad t > 90$

(i) Sketch the speed–time graph for Christine's descent and explain the shape of each section.

(ii) How high is the aircraft when Christine jumps out?

(iii) Write down expressions for the acceleration during the various phases of Christine's descent. What is the greatest magnitude of her acceleration?

(iv) Do you think this model is realistic?

10. A train starts from rest at a station. Its acceleration is shown on the acceleration–time graph below.

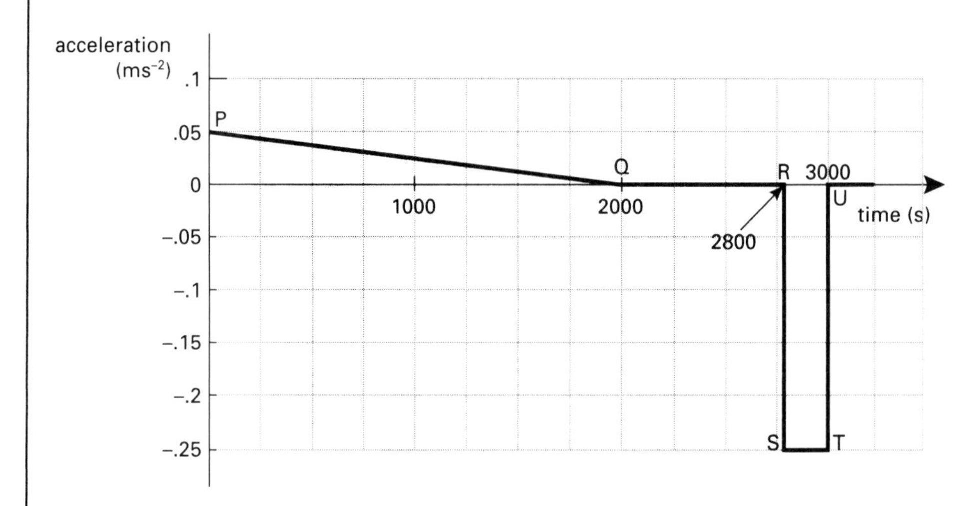

(i) Describe what is happening during the phases of the train's journey represented by the lines PQ, QR and ST.

(ii) The equation of the line PQ is of the form $a = mt + c$. Find the values of the constants m and c.

(iii) Find the maximum speed of the train.

(iv) What is the speed of the train when $t = 3000$ seconds?

(v) How far does the train travel during the first 3000 seconds?

For Discussion

Which of the following do you think could be modelled as motion in a straight line with constant acceleration?

(i) a ball thrown horizontally from a tower;
(ii) a coin tossed vertically into the air to see how high it goes;
(iii) a car travelling along a straight road at 60 mph;
(iv) a skier with acceleration 6 ms^{-2} on a smooth ski-slope;
(v) a 110 m hurdler throughout a race.

Equations of constant acceleration

Consider the motion of an object in a straight line with constant acceleration a. If the position of the object relative to its starting point is s then $\frac{ds}{dt}$ is its velocity and $\frac{d^2s}{dt^2}$ is one way of expressing its acceleration.

Suppose that initially the object's velocity is u.

Since $a = \frac{dv}{dt}$ we can integrate to give $v = at + u$.

This is one of the equations of constant acceleration that you will find useful in many problems. It is more usually written

$$v = u + at.$$

Since $v = \frac{ds}{dt}$ we can integrate again to give

$$s = ut + \tfrac{1}{2}at^2 + s_0.$$

It is usual to take the origin to be the initial position of the particle and so s_0 to be zero, giving a second equation of constant acceleration:

$$s = ut + \tfrac{1}{2}at^2.$$

Eliminating t between the two equations gives a third equation

$$v^2 = u^2 + 2as$$

and eliminating a gives a fourth equation

$$s = \tfrac{1}{2}(u + v)t.$$

So when an object moves *along a straight line with constant acceleration*, its velocity v and displacement s (from its initial position) at time t are given by the equations

$$v = u + at$$

$$v^2 = u^2 + 2as$$

$$s = ut + \tfrac{1}{2}at^2$$

$$s = \tfrac{1}{2}(u + v)t$$

where u is the initial velocity of the object.

EXAMPLE

A stone is dropped from rest at the top of a building of height 12 metres and travels in a straight line with constant acceleration 10 ms^{-2}. Find the time it takes to reach the ground and the speed of impact.

Solution

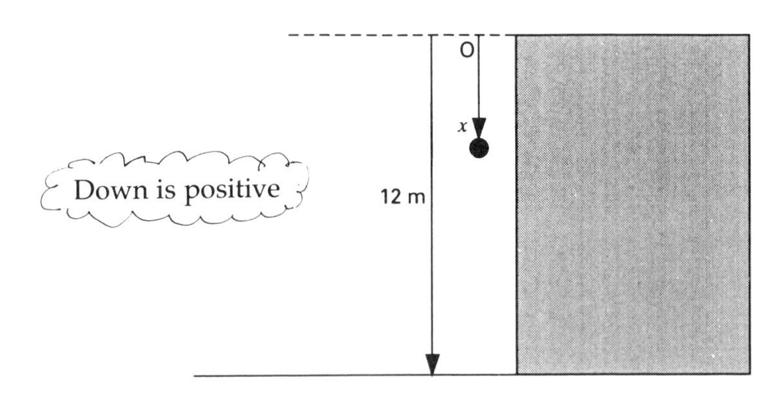

We are told that the acceleration is constant so we can use the constant acceleration formulae, with $a = 10$ and $u = 0$.

$$s = ut + \tfrac{1}{2}at^2 \Rightarrow s = 0 + 5t^2$$

When the stone hits the ground $s = 12$ m so

$$t^2 = \frac{12}{5} = 2.4$$

$$t = 1.55 \text{ seconds}$$

$$v^2 = u^2 + 2as \Rightarrow v^2 = 0 + 20 \times 12$$

$$v^2 = 240$$

$$v = 15.5 \text{ ms}^{-1}$$

The stone takes 1.55 s to hit the ground and has speed 15.5 ms^{-1} on impact.

Note that in solving this problem we have used only two of the four constant acceleration equations.

A ball is thrown vertically up in the air with initial speed 5 ms^{-1}. It has constant acceleration 10 ms^{-2} vertically downwards due to gravity. Find the maximum height reached, and how long it takes to reach it.

Solution

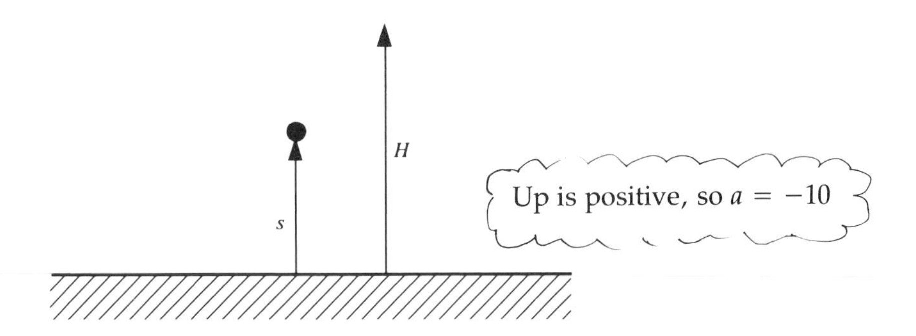

The acceleration given is constant. $a = -10$ and $u = +5$.

At the point of maximum height, let $s = H$. The ball instantaneously stops before falling so at this point $v = 0$. From equation (2)

$$v^2 = u^2 + 2as$$

$$0 = 25 - 20H$$

$$\Rightarrow H = 1.25 \text{ m}$$

The maximum height of the ball above the point of release is 1.25 metres.

At maximum height $v = 0$, so $v = u + at$

$$\Rightarrow 0 = 5 - 10t$$

$$t = 0.5$$

The ball takes half a second to reach its maximum height.

Exercise 2F

1. A stone is thrown vertically upwards with an initial speed of 25 ms^{-1} towards a balloon. Its acceleration is 10 ms^{-2} directed vertically downwards. If the stone hits its target after 2 seconds find its speed at that point, and the height of the balloon.

Exercise 2F continued

2. A car starting from rest at traffic lights reaches a speed of 90 kmh^{-1} in 12 seconds. Find the acceleration of the car (in ms^{-2}) and the distance travelled. Write down any assumptions that you have made.

3. Use the equations of constant acceleration to complete the following table.

displacement (m)	velocity (ms^{-1}) initial	final	time (s)	acceleration (ms^{-2})
	24	0		−0.8
	10	40	10	
	0		4	4
16		0		−8
10	4		2	

4. A top sprinter accelerates from rest to 9 ms^{-1} in 2 seconds. Calculate his acceleration, assumed constant, during this period.

5. A van skids to a halt from an initial speed of 24 ms^{-1}, covering a distance of 36 m. Find the acceleration of the van (assumed constant) and the time it takes to stop.

6. A car is travelling along a straight road. It accelerates uniformly from rest to a speed of 15 ms^{-1} and maintains this speed for 10 minutes. It then decelerates uniformly to rest. If the acceleration and deceleration are 5 ms^{-2} and 8 ms^{-2} respectively, find the total journey time and the total distance travelled during the journey.

7. A skier pushes off at the top of a slope with an initial speed of 2 ms^{-1}. She gains speed at a constant rate throughout her run. After 10 s she is moving at 6 ms^{-1}.
   (i) Find an expression for her speed t seconds after she pushes off.
   (ii) Find an expression for the distance she has travelled at time t seconds.
   (iii) The length of the ski slope is 400 m. What is her speed at the bottom of the slope?

8. Towards the end of a half-marathon Sabina is 100 metres from the finish line and is running at a constant speed of 5 ms^{-1}. Daniel, who is 140 metres from the finish and is running at 4 ms^{-1}, decides to accelerate to try to beat Sabina. If he accelerates uniformly at 0.25 ms^{-2}, does he succeed?

9. A ball is dropped from a building of height 30 metres, and at the same instant a stone is thrown vertically upwards from the ground so that it hits the ball. In modelling the motion of the ball and stone it is assumed that each object moves in a straight line with a constant

downward acceleration of magnitude 10 ms^{-2}. The stone is thrown with initial speed of 15 ms^{-1} and is h metres above the ground t seconds later.

(i) Draw a diagram of the ball and stone before they collide, marking their positions.
(ii) Write down an expression for h at time t.
(iii) Write down an expression for the height of the ball at time t.
(iv) When do the ball and stone collide?
(v) How high above the ground do the ball and stone collide?

10. The speed of an accelerating car at various times is shown in the table below.

time (s)	1	2	3	4	5	6	7	8	9	10
speed (ms^{-1})	6.7	9.5	9.4	12.2	15.0	17.5	17.4	20.1	23.0	25.8

(i) Draw the speed–time graph for the car.
(ii) Estimate the initial speed of the car.
(iii) When accelerating, the car driver makes two gear changes. During each gear change, which lasts for 1 second, the car slows down with a constant deceleration. When do the gear changes occur, and what is the size of the deceleration?
(iv) Between gear-changes, does the car have uniform acceleration? If so, what is its magnitude?

11. When Kim rows her boat, the 2 oars are both in the water for 3 seconds and then both out of the water for 2 seconds. This 5 second cycle is then repeated.

When the oars are in the water the boat accelerates at a constant 1.8 ms^{-2}, and when they are not in the water it decelerates at a constant 2.2 ms^{-2}.

(i) Find the change in speed that takes place in each 3 second period of acceleration.
(ii) Find the change in speed that takes place in each 2 second period of deceleration.
(iii) Calculate the change in the boat's speed for each 5 second cycle.
(iv) A race takes Kim 45 seconds to complete. If she starts from rest what is her speed as she crosses the finishing line?
(v) Discuss whether this is a realistic speed for a rowing boat.

12. A ball is dropped from a tall building and falls with acceleration of magnitude 10 ms^{-2}. The distance between floors in the block is constant. The ball takes 1 s to fall from the 14th to the 13th floor and 0.5 s to fall from the 13th floor to the 12th. What is the distance between floors?

Exercise 2F continued

13. Two clay pigeons are launched vertically upwards from exactly the same spot at 1 second intervals. Each pigeon has initial speed 30 ms^{-1} and acceleration 10 ms^{-2} downwards. How high above the ground do they collide?

Investigations

The Two Second Rule

The 'Two Second Rule' states that if you are driving a car you should be at least 2 seconds behind the car in front. It provides a quick check that you are leaving yourself a safe stopping distance. Many drivers say the phrase 'Only a fool breaks the Two Second Rule' (which takes about 2 seconds to say) to check that they are a safe distance behind the car in front.

Investigate whether the Two Second Rule is a good guide to safe driving.

Convoy

You may have noticed that when vehicles are moving in convoy, such as in a funeral procession, there can be a significant variation in speed of the vehicles at the end of the convoy even if the leading vehicle is driven at a fairly close approximation to constant speed.

Investigate the effect on a convoy of vehicles of variations in the speed of the first vehicle.

Suggest recommendations for the speed and maximum length of a convoy.

Train journey

If you look out of a train window you will see distance markers beside the track, every quarter of a mile. Take a train journey and record the time as you go past each of these markers. Use your figures to draw distance–time, speed–time and acceleration–time graphs.

What can you conclude about the greatest acceleration, deceleration and speed of the train?

Speed bumps

The residents of a housing estate are worried about the danger from cars being driven around at high speed, and they request that speed bumps be installed.

How far apart should the bumps be placed?

KEY POINTS

- **Position** is given relative to the origin.
- **Displacement** is change in position from a given point.
- **Distance travelled** is the amount of ground (etc.) covered.
- **Speed** has value but no specified direction.
- **Average speed** $= \dfrac{\text{distance travelled}}{\text{time taken}}$.
- **Velocity** is speed in a stated direction.
- **Kinematics Graphs.**

	Position–time graph	Velocity–time graph	Acceleration–time graph
Gradient	velocity	acceleration	nothing useful
Area under graph	nothing useful	displacement	velocity

- Using calculus, position, velocity and acceleration are related by

	Differentiation (↓)	Integration (↑)
Position	s	$s = \int v\,dt$
Velocity	$v = \dfrac{ds}{dt}$	$v = \int a\,dt$
Acceleration	$a = \dfrac{dv}{dt} = \dfrac{d^2s}{dt^2}$	a

- **Constant acceleration formulae:**

$$v = u + at$$

$$s = ut + \tfrac{1}{2}at^2$$

$$s = \tfrac{1}{2}(u + v)t$$

$$v^2 = u^2 + 2as$$

3 Vectors

But the principal failing occurred in the sailing
And the bellman, perplexed and distressed,
Said he had hoped, at least when the wind blew due East
That the ship would not travel due West.

Lewis Carroll

Suppose you have followed these instructions:
walk 4 km north-east and then 3 km east.

(i) How can you record what you have done?
(ii) How far have you walked?
(iii) In what direction have you walked?

In this situation each instruction has two parts: a distance, like 4 km. and a direction, north-east. Quantities like this that have both size and direction are called *vectors*.

You will meet many vector quantities in mechanics: displacement, velocity, acceleration and force, for example, are vector quantities. All vectors share the same properties, whatever they represent. Many of the techniques in this chapter are introduced in the context of displacements, but they would apply equally to any other vector quantity.

Displacement vectors

The instruction 'walk 4 km north-east and then 3 km east' can be modelled mathematically using a scale diagram, as in figure 3.1. The arrowed lines AB and BC are examples of vectors.

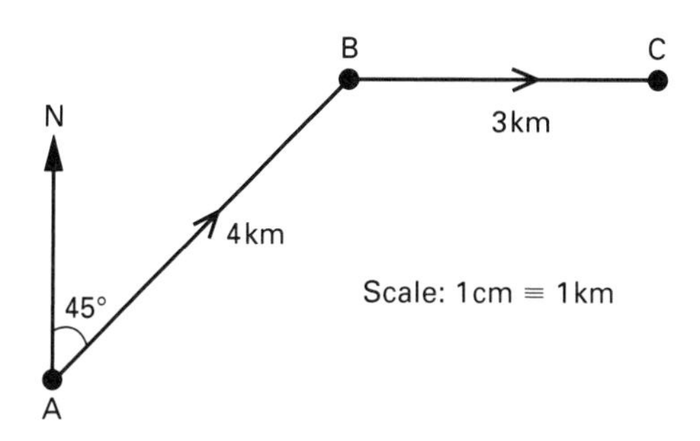

Figure 3.1

We write the vectors as $\overrightarrow{AB}$ and $\overrightarrow{BC}$: the arrows above the letters are very important as they indicate the directions of the vectors. $\overrightarrow{AB}$ means from A to B. $\overrightarrow{AB}$ and $\overrightarrow{BC}$ are examples of *displacement vectors*.

It is often more convenient to use a single letter to denote a vector. For example in textbooks you might see the displacement vector $\overrightarrow{AB}$ written as **p** and $\overrightarrow{BC}$ as **q** (i.e. in bold print). When writing vectors yourself, you should underline your letters e.g. $\underline{p}$ and $\underline{q}$.

Position vectors

The first step in modelling in mechanics is often to represent objects as particles or points moving in space. To specify the position of an object you define its displacement relative to a fixed origin.

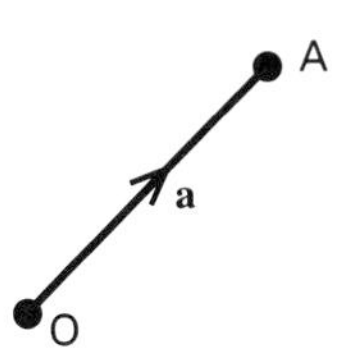

Figure 3.2

In figure 3.2 the vector **a** (or $\overrightarrow{OA}$) is called the *position vector* of A. It defines uniquely the position of the point A. This is the difference between a position vector and a displacement vector. 'Twenty metres north' is a displacement vector. 'Twenty metres north of Nelson's Column' is a position vector. Both types are useful in different situations.

Magnitude and direction

A vector is defined by its magnitude (size) and direction. The magnitude of a vector **a** is written either |**a**| or a (i.e. in ordinary print).

There are two common ways of giving direction.

(i) **The mathematicians' convention**
In this method the reference direction is the x axis; angles above the x axis are positive and angles below the x axis are negative (figure 3.3).

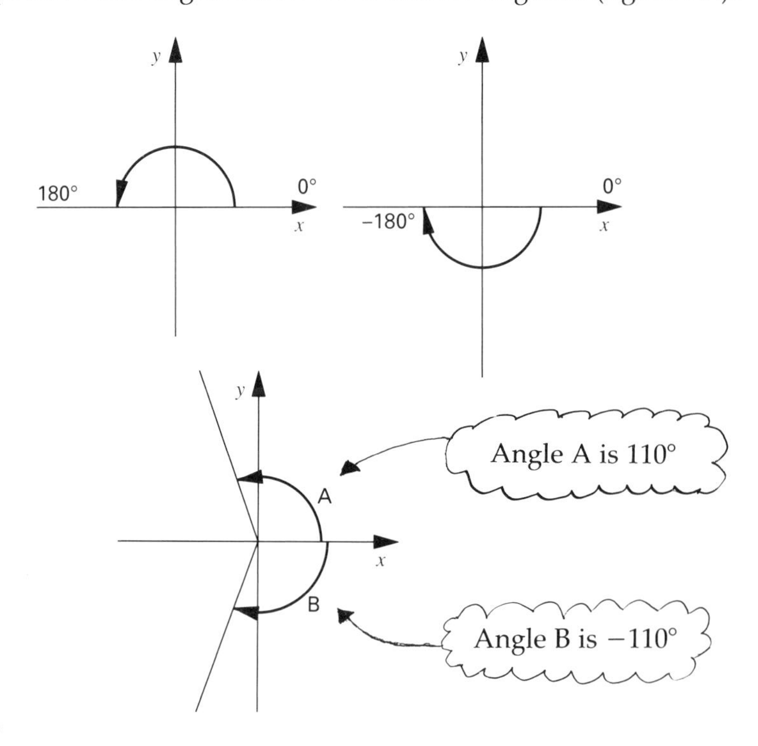

Figure 3.3

(ii) **Compass bearings**
In this method the reference direction is due north and the angle is measured clockwise (figure 3.4).

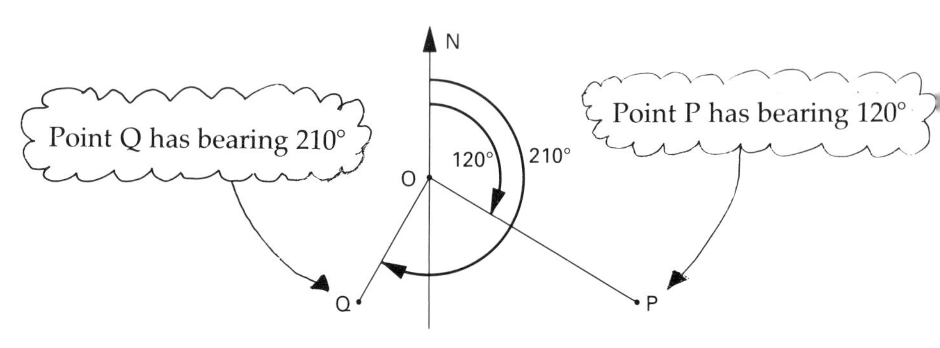

Figure 3.4

Addition of vectors

Consider the two displacements in the example at the start of this chapter: 'walk 4 km north-east, and then 3 km due east.'

Labelling the two displacements as $\overrightarrow{AB}$ and $\overrightarrow{BC}$, then the combined effect of these displacements is $\overrightarrow{AC}$ (as in figure 3.5). We refer to this process of combining vectors as vector addition and the outcome of doing it as the *resultant vector*.

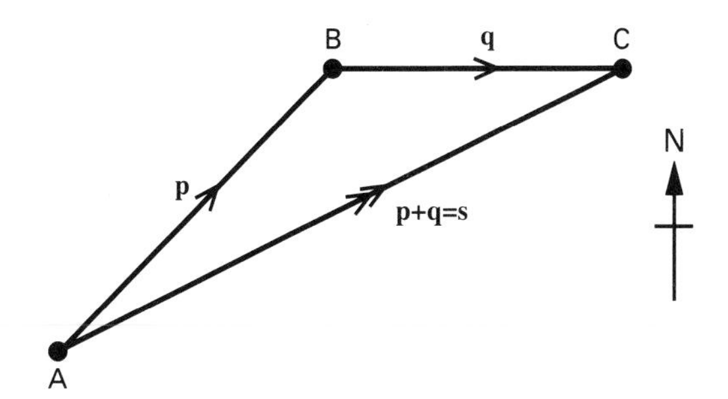

Figure 3.5

We write $\overrightarrow{AB} + \overrightarrow{BC} = \overrightarrow{AC}$ or $\mathbf{p} + \mathbf{q} = \mathbf{s}$.

This leads us naturally to the law of vector addition.

To add several vectors, we draw them end-to-end, starting one vector at the end of the previous one, as in figure 3.6. The resultant is the vector which goes from the start of the first vector to the end of the last one. It is marked with two arrows.

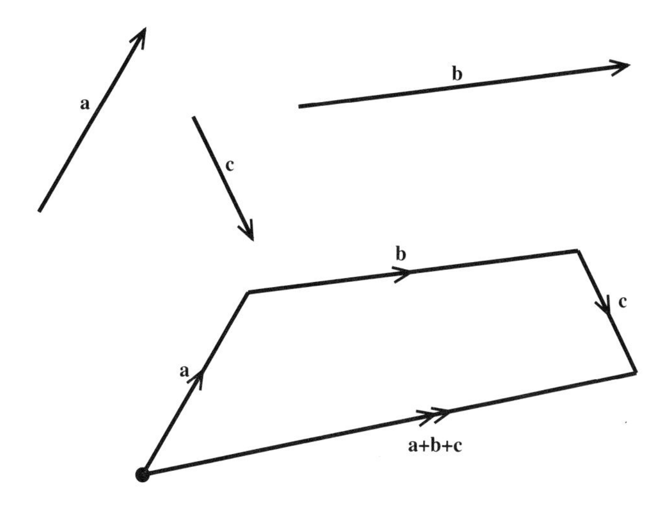

Figure 3.6

EXAMPLE

Consider the three vectors **a**, **b** and **c** shown in the xy plane. Draw scale diagrams to show the addition of vectors (i) $\mathbf{a} + \mathbf{b}$, (ii) $\mathbf{b} + \mathbf{c}$ and (iii) $\mathbf{a} + \mathbf{b} + \mathbf{c}$.

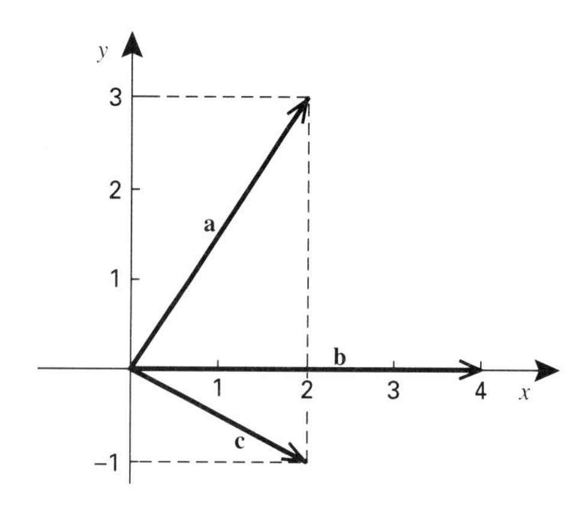

Solution

Using the law of vector addition the solutions are as shown below.

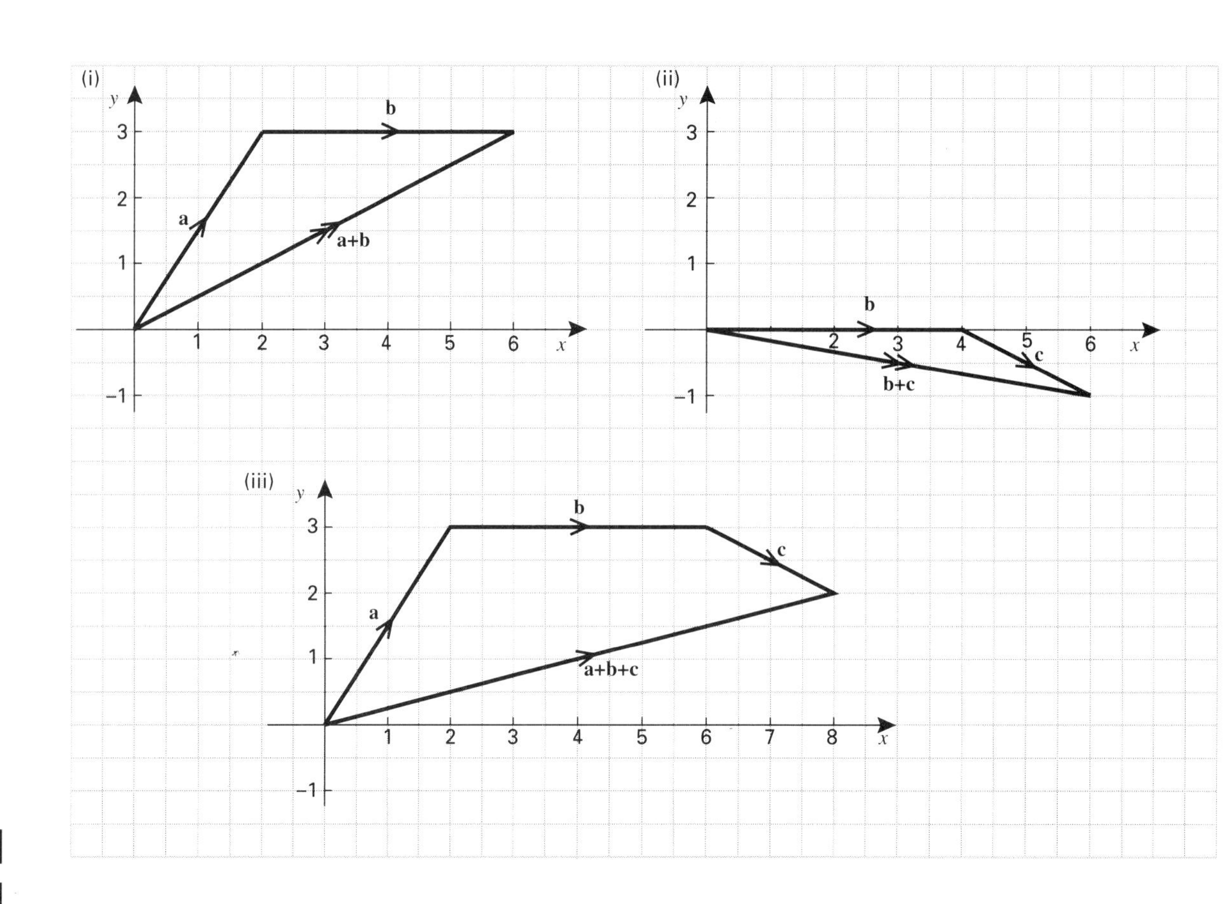

The first of our instructions at the start of the chapter was 'walk 4 km north-east'. Suppose that we wanted to retrace our steps. Obviously we should have to walk 4 km south-west, the same distance but in the opposite direction. If our first instruction is represented by the vector $\overrightarrow{AB}$ then the instruction to go back would be $\overrightarrow{BA}$. If we write $\overrightarrow{AB} = \mathbf{a}$ then we write $\overrightarrow{BA} = -\mathbf{a}$.

You will often see expressions like $\mathbf{a} - \mathbf{b}$. This is a shorthand way of writing $\mathbf{a} + (-\mathbf{b})$.

If we add the vectors $\overrightarrow{AB}$ and $\overrightarrow{BA}$ we obtain a resultant with zero length, which is an example of the zero vector. We would write

$$\mathbf{a} + (-\mathbf{a}) = \mathbf{0}.$$

Note that we use $\mathbf{0}$ and not 0 on the right-hand side of this expression to show that the quantity is still a vector.

EXAMPLE Given the vectors **a** and **b** as shown, draw the vector $\mathbf{a} - \mathbf{b}$.

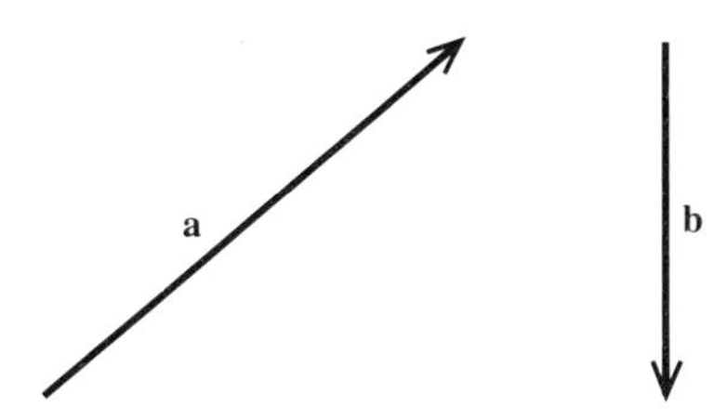

Solution
In order to draw $\mathbf{a} - \mathbf{b}$ we first have to create the vector $(-\mathbf{b})$ so that we can carry out the addition $\mathbf{a} + (-\mathbf{b})$. This is the same as **b** but in the opposite direction.

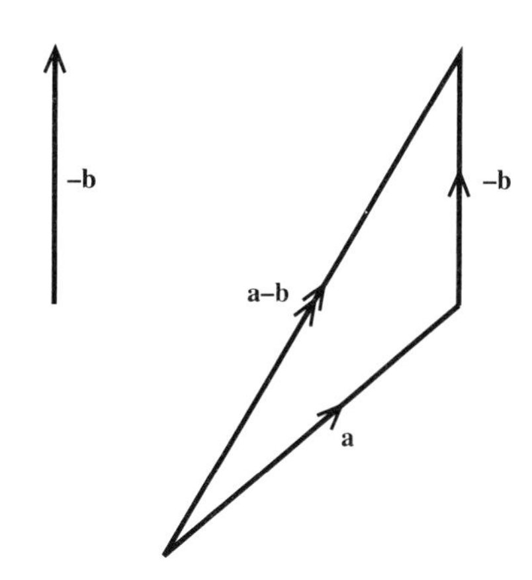

Multiplying by a scalar

If you carry out the instruction 'walk 4 km north-east' three times, then you walk twelve km north-east. In general, if n is a positive number, adding n lots of the vector **a** will give a vector with n times the magnitude of **a** in the same direction as **a**. We write this vector as n**a** and call it a *scalar multiple* of **a**. The use of the term scalar emphasises that n is not a vector: it has size but no direction.

EXAMPLE

The figure shows two vectors **a** and **b** with magnitudes 2 cm and 3 cm.

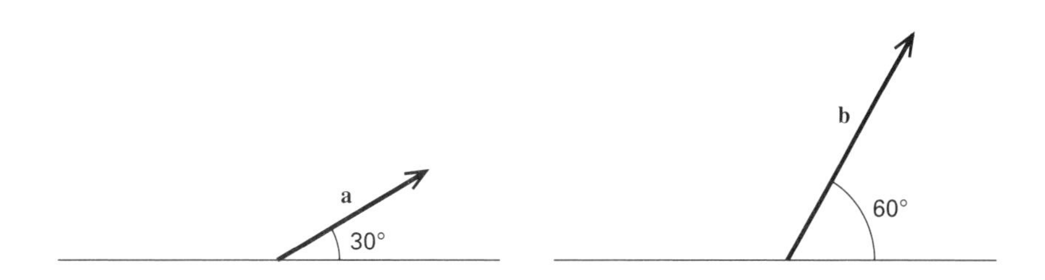

(i) Draw scale diagrams to show the vectors 2**a**, 1.5**b** and −2**b**.
(ii) Construct the vectors 2**a** + 1.5**b** and 2**a** − 2**b**.

Solution

(i)

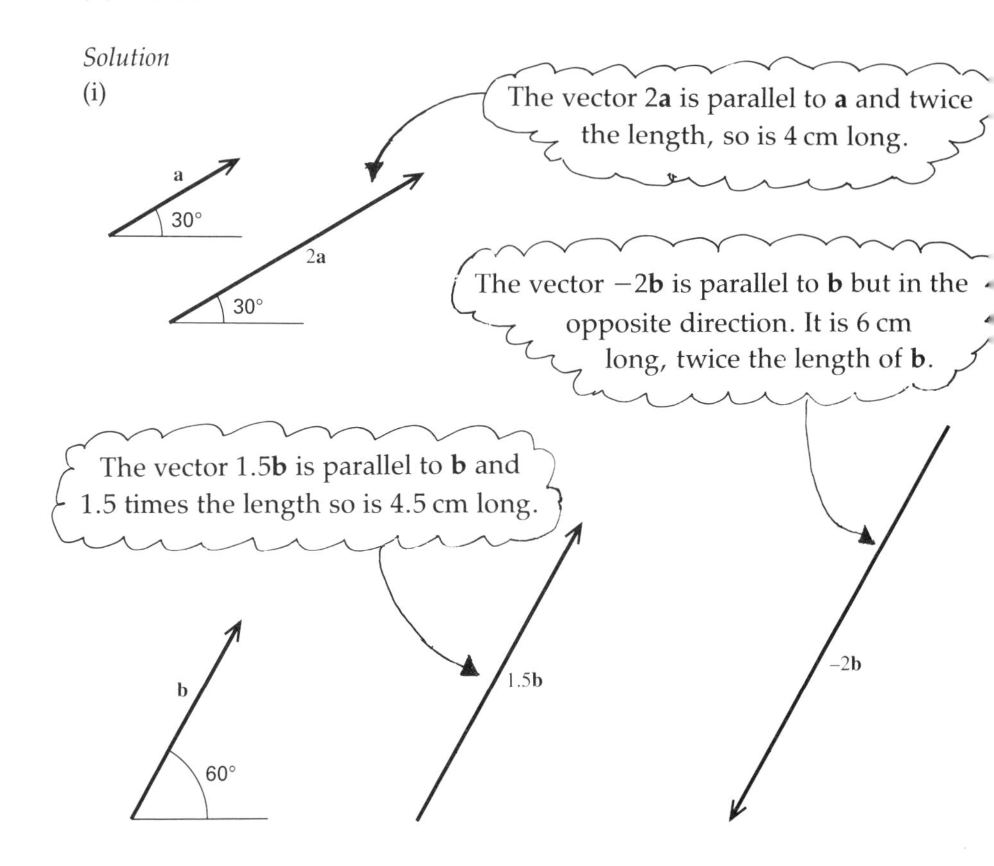

(ii) Using the law of vector addition gives the following diagrams.

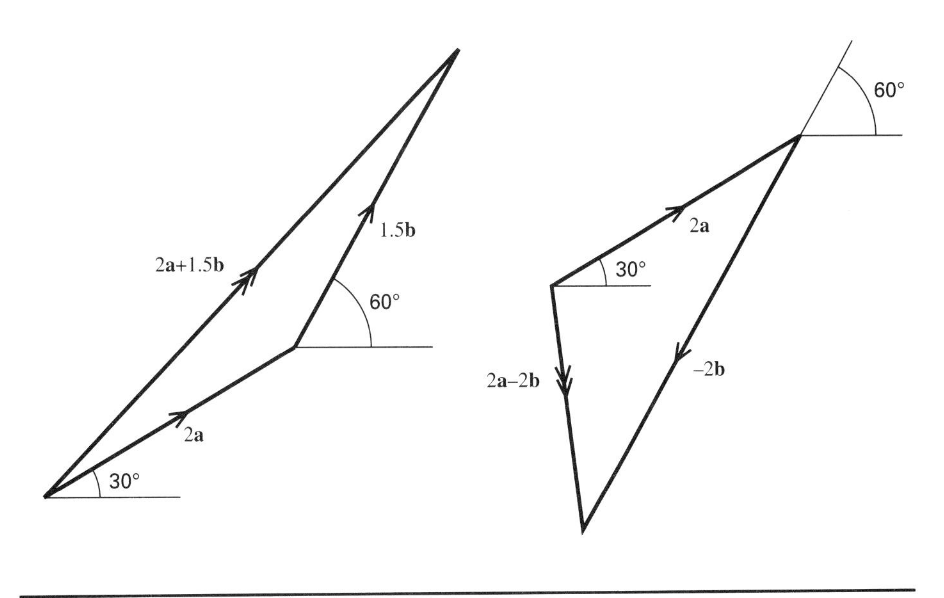

Exercise 3A

1. The diagram shows the relative positions of various towns in England. Show the following displacements and describe them in words:

(i) Cambridge from Wolverhampton,
(ii) Wolverhampton from Cambridge,
(iii) Reading from Wolverhampton,
(iv) Cambridge from Reading.

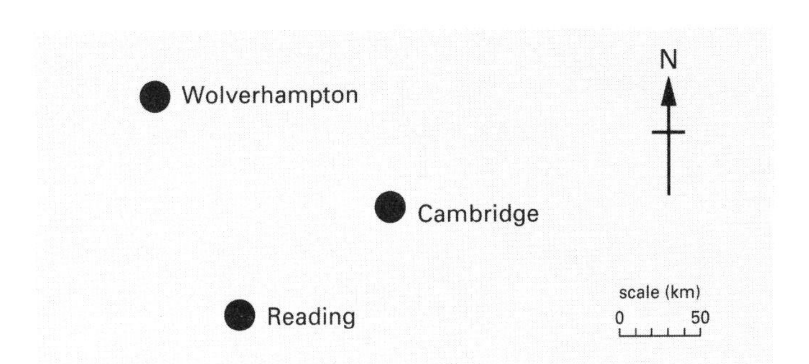

2. The figure shows several vectors. Which vectors are equal to $\mathbf{a}$, $\mathbf{c}$ and $-\mathbf{a}$?

Exercise 3A continued

3. A child climbs up the ladder attached to a slide and then slides down. What three vectors model the displacement of the child during this activity?

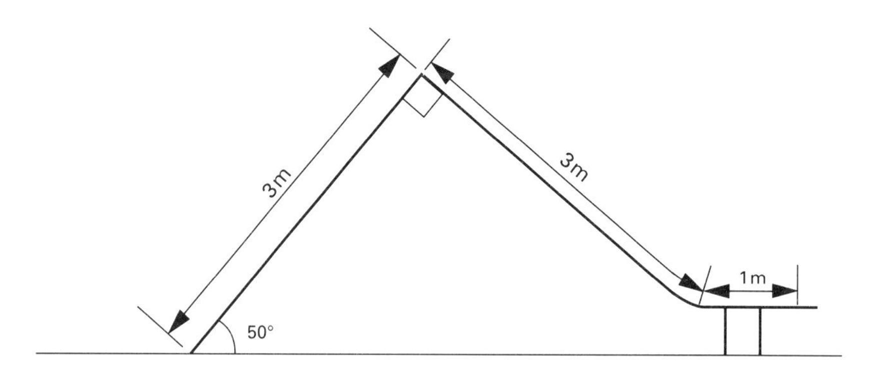

4. A yacht travels 6 km on a bearing 030° and then 3 km due south.
   (i) Draw a scale diagram to show the path of the yacht.
   (ii) How far is it from its starting point, and in what direction?

5. A frigate sails 10 km due north and then 8 km due east. Draw a scale diagram to show the addition of these displacements. Describe the resultant of the journey.

6. A crane moves a crate from the ground 10 m vertically upward, then 6 m horizontally and 2 m vertically downward. Draw a scale diagram of the path of the crate. What single translation would move the crate to its final position from its initial position on the ground?

7. A boy walks 30 m north and then 50 m south-west.
   (i) Draw a diagram to show the boy's path.
   (ii) In which direction should the boy walk to get directly back to his starting point?

8. The three vectors $\mathbf{a}$, $\mathbf{b}$ and $\mathbf{c}$ have magnitudes 1, 2 and 5 as shown in the diagram.

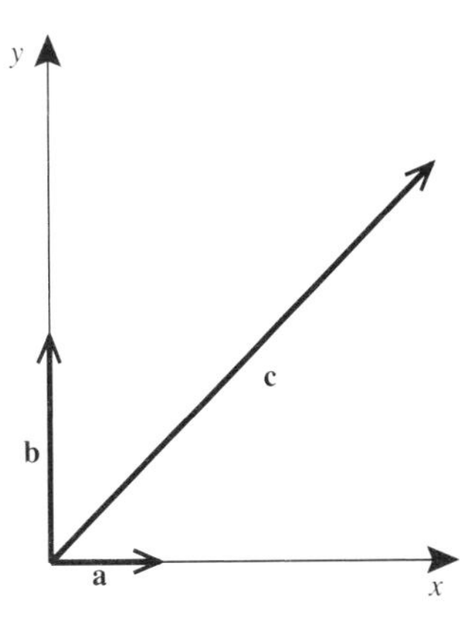

(i) Draw scale diagrams to illustrate $\mathbf{a} + \mathbf{b}$ and $\mathbf{b} + \mathbf{a}$. What do you conclude?

(ii) Draw a scale diagram to illustrate $\mathbf{a} + \mathbf{b} + \mathbf{c}$. Explain why you need do no more work to illustrate $\mathbf{c} + \mathbf{a} + \mathbf{b}$.

9. The diagram shows an irregular pentagon drawn to scale.

Write down relationships between the following vectors:

(i) $\overrightarrow{AB}$ and $\overrightarrow{DC}$, (ii) $\overrightarrow{AB}$ and $\overrightarrow{CD}$,

(iii) $\overrightarrow{EA}$ and $\overrightarrow{CB}$, (iv) $\overrightarrow{AE}$ and $\overrightarrow{CB}$.

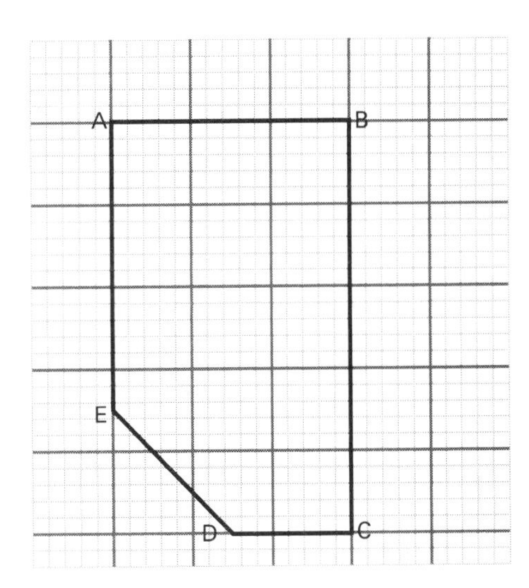

10. ABCD is a quadrilateral. Show that

$$\overrightarrow{AB} + \overrightarrow{BC} + \overrightarrow{CD} + \overrightarrow{DA} = \mathbf{0}.$$

11. The triangle ABC is such that $\overrightarrow{AB} = \mathbf{p}$ and $\overrightarrow{AC} = \mathbf{q}$.

Find expressions in terms of $\mathbf{p}$ and $\mathbf{q}$ for:

(i) $\overrightarrow{BC}$ (ii) $\overrightarrow{BP}$ where $\overrightarrow{BP} = \frac{1}{3}\overrightarrow{BC}$

(iii) $\overrightarrow{AP}$ (iv) $\overrightarrow{AQ}$ where Q is the midpoint of AP.

Exercise 3A continued

12. In the parallelogram below, $\overrightarrow{OA} = \mathbf{a}$, $\overrightarrow{OC} = \mathbf{c}$, and M is the midpoint of AB. Express the displacements below in terms of $\mathbf{a}$ and $\mathbf{c}$:

(i) $\overrightarrow{OB}$ (ii) $\overrightarrow{AC}$ (iii) $\overrightarrow{CA}$ (iv) $\overrightarrow{BO}$
(v) $\overrightarrow{AM}$ (vi) $\overrightarrow{OM}$ (vii) $\overrightarrow{MC}$

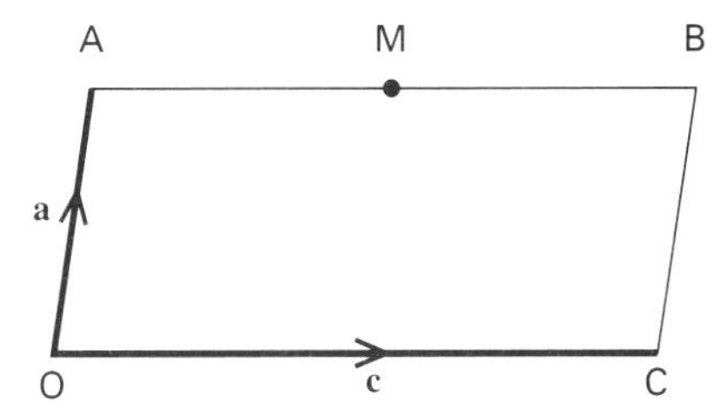

13. Consider the regular hexagon shown below. Given that $\overrightarrow{GB} = \mathbf{p}$ and $\overrightarrow{GC} = \mathbf{q}$, express the following in terms of $\mathbf{p}$ and $\mathbf{q}$:

(i) $\overrightarrow{BC}$ (ii) $\overrightarrow{CA}$ (iii) $\overrightarrow{GA}$ (iv) $\overrightarrow{CD}$ (v) $\overrightarrow{GE}$ (vi) $\overrightarrow{GF}$

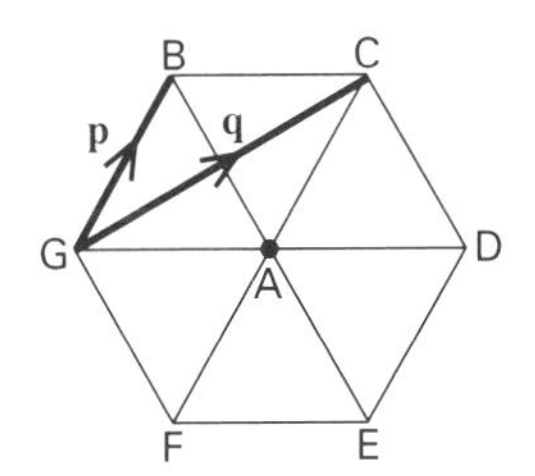

14. A sailing boat cannot sail closer than 45° to the direction from which the wind is blowing. The boat has to sail from A to B. What route would you advise the helmsman to take in order to get there as quickly as possible?

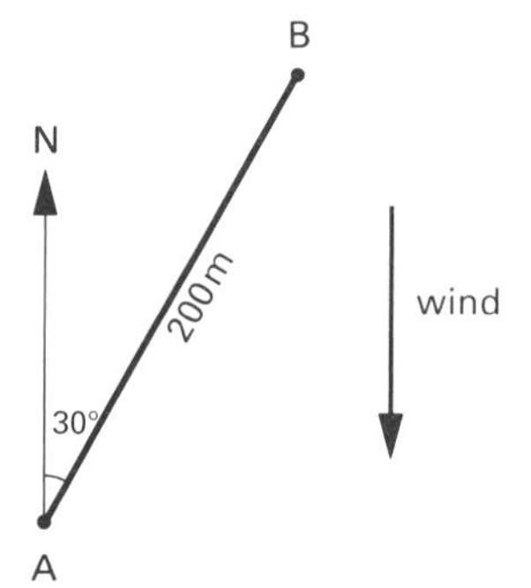

Scale drawing and trigonometry

Diagrams showing the addition (etc.) of vectors may be used in two ways to find the magnitude and direction of the resultant:

(a) by accurate scale drawing and measurement of the required lengths and angles;

(b) by calculation, using trigonometry on the diagram.

The use of a scale drawing often leads to a quick answer, but it will only be approximate.

EXAMPLE

Consider the instruction 'Walk 4 km north-east and then 3 km east'. Use trigonometry to calculate the length and direction of the resultant displacement, correct to two significant figures.

Solution
First we sketch a diagram, marking on all the available information.

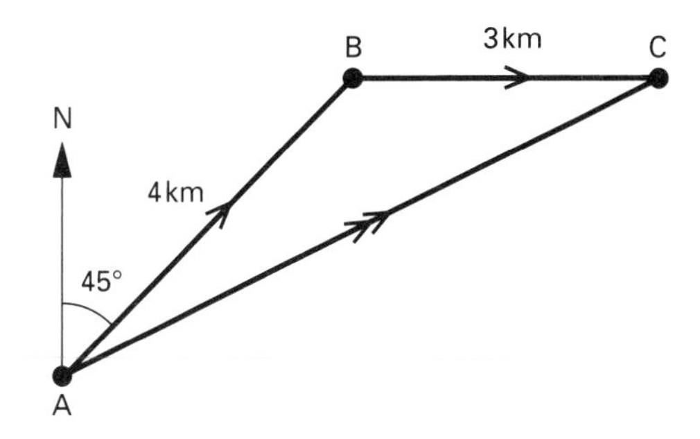

Clearly angle B is 135° and we can use the cosine rule in triangle ABC to find AC.

$$
\begin{aligned}
AC^2 &= 4^2 + 3^2 - 2 \times 4 \times 3 \times \cos 135^\circ \\
&= 41.9706 \\
AC &= 6.4784
\end{aligned}
$$

The distance is 6.5 km (to two significant figures).

Using the sine rule on triangle ABC: $\sin A = \dfrac{\sin 135^\circ}{6.4784}$

Solving for A gives A = 19.11°, which is 19° (to two significant figures).

The bearing of AC is 19° + 45° = 64°.

Components of a vector

So far we have explored the properties of vectors by drawing lines to represent displacements. An alternative approach involves splitting vectors into *components* so that we specify the distance travelled in fixed directions. For instance, a journey could be specified by saying that it involved a displacement of 4 km north as well as 3 km east, as shown in figure 3.7.

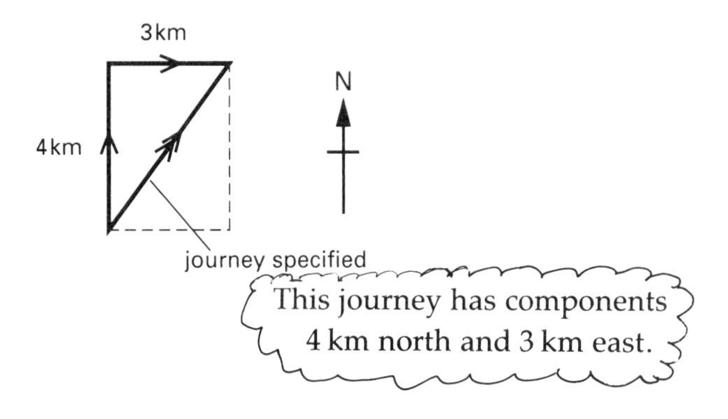

Figure 3.7

We can write this as $3\mathbf{i} + 4\mathbf{j}$ where $\mathbf{i}$ represents a displacement of 1 km to the east, and $\mathbf{j}$ 1 km to the north.

Alternatively it can be written $\begin{pmatrix} 3 \\ 4 \end{pmatrix}$; this is called a *column vector*.

We often find it convenient to relate vectors to the standard Cartesian coordinate system. In this case we denote a vector of one unit along the x axis as $\mathbf{i}$ and a vector of one unit along the y axis as $\mathbf{j}$. Any other vector drawn in the xy plane can then be written in terms of $\mathbf{i}$ and $\mathbf{j}$.

You may define the unit vectors $\mathbf{i}$ and $\mathbf{j}$ to be in other directions if it is convenient to do so.

EXAMPLE

Write the four vectors $\mathbf{a}$, $\mathbf{b}$, $\mathbf{c}$ and $\mathbf{d}$ shown in the diagram in component form.

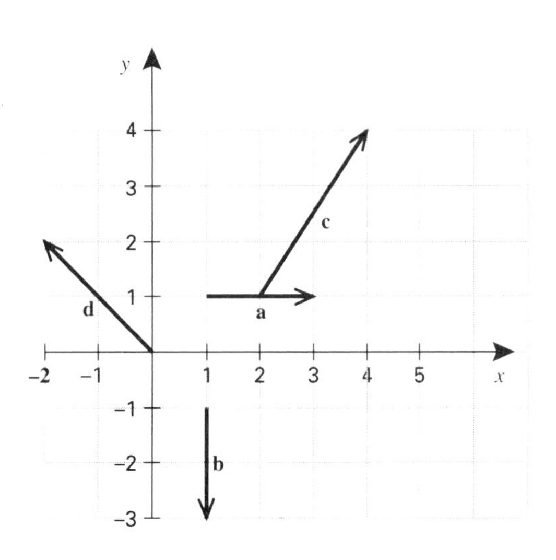

Solution

$$\mathbf{a} = 2\mathbf{i} = \begin{pmatrix} 2 \\ 0 \end{pmatrix}$$

$$\mathbf{b} = -2\mathbf{j} = \begin{pmatrix} 0 \\ -2 \end{pmatrix}$$

$$\mathbf{c} = 2\mathbf{i} + 3\mathbf{j} = \begin{pmatrix} 2 \\ 3 \end{pmatrix}$$

$$\mathbf{d} = -2\mathbf{i} + 2\mathbf{j} = \begin{pmatrix} -2 \\ 2 \end{pmatrix}$$

Equality of vectors

If two vectors **p** and **q** are equal then they must be equal in both magnitude and direction. If they are written in component form their components must be equal.

So if $\mathbf{p} = a_1\mathbf{i} + b_1\mathbf{j}$
and $\mathbf{q} = a_2\mathbf{i} + b_2\mathbf{j}$,
then $a_1 = a_2$ and $b_1 = b_2$.

Thus in two dimensions, the statement $\mathbf{p} = \mathbf{q}$ is the equivalent of two equations (and in three dimensions, three equations).

Addition of vectors in component form

In component form, addition and subtraction of vectors is simply carried out by adding or subtracting the components of the vectors.

EXAMPLE

Two vectors **a** and **b** are given by $\mathbf{a} = 2\mathbf{i} + 3\mathbf{j}$ and $\mathbf{b} = -\mathbf{i} + 4\mathbf{j}$.

(i) Find the vectors $\mathbf{a} + \mathbf{b}$ and $\mathbf{a} - \mathbf{b}$.

(ii) Verify that your results are the same if you use a scale drawing.

Solution

(i) Using **i** and **j**:

$$\begin{aligned}\mathbf{a} + \mathbf{b} &= (2\mathbf{i} + 3\mathbf{j}) + (-\mathbf{i} + 4\mathbf{j}) \\ &= 2\mathbf{i} - \mathbf{i} + 3\mathbf{j} + 4\mathbf{j} \\ &= \mathbf{i} + 7\mathbf{j}\end{aligned}$$

$$\begin{aligned}\mathbf{a} - \mathbf{b} &= (2\mathbf{i} + 3\mathbf{j}) - (-\mathbf{i} + 4\mathbf{j}) \\ &= 2\mathbf{i} + \mathbf{i} + 3\mathbf{j} - 4\mathbf{j} \\ &= 3\mathbf{i} - \mathbf{j}\end{aligned}$$

Using column vectors:

$$\mathbf{a} + \mathbf{b} = \begin{pmatrix} 2 \\ 3 \end{pmatrix} + \begin{pmatrix} -1 \\ 4 \end{pmatrix} = \begin{pmatrix} 1 \\ 7 \end{pmatrix}$$

$$\mathbf{a} - \mathbf{b} = \begin{pmatrix} 2 \\ 3 \end{pmatrix} - \begin{pmatrix} -1 \\ 4 \end{pmatrix} = \begin{pmatrix} 3 \\ -1 \end{pmatrix}$$

(ii)

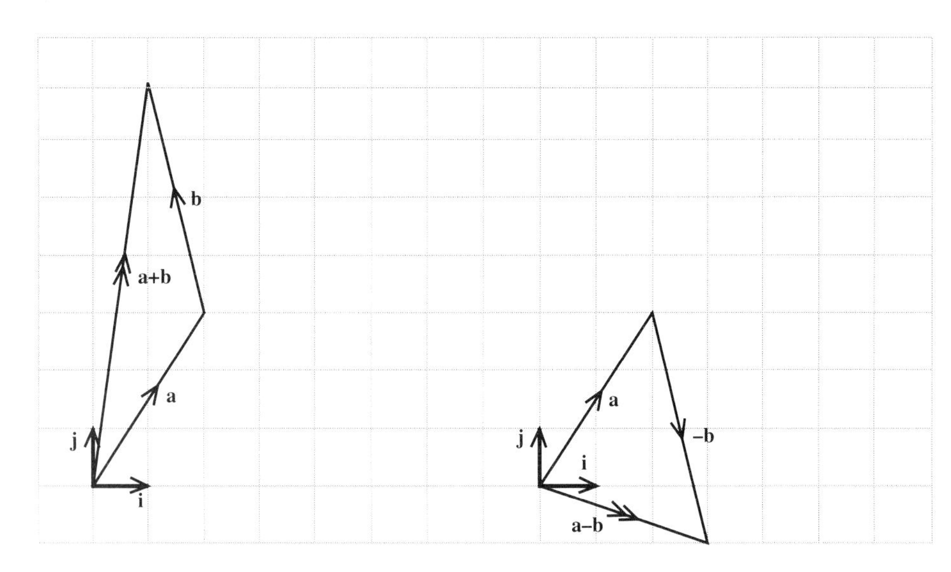

From the diagram you can see that $\mathbf{a} + \mathbf{b} = \mathbf{i} + 7\mathbf{j}$ or $\begin{pmatrix} 1 \\ 7 \end{pmatrix}$

and $\mathbf{a} - \mathbf{b} = 3\mathbf{i} - \mathbf{j}$ or $\begin{pmatrix} 3 \\ -1 \end{pmatrix}$.

These vectors are the same as those obtained in part (i).

Exercise 3B

1. The diagram shows a grid of 1 metre squares. A person walks first east and then north. How far should the person walk in each of these directions to travel
(i) from A to B, (ii) from C to D, (iii) from A to D?

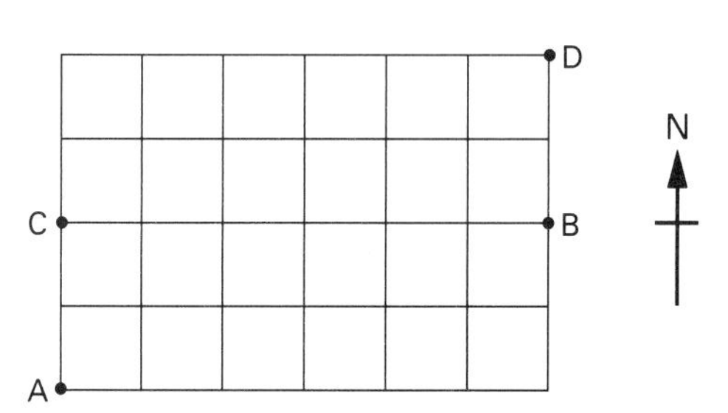

2. Write the vectors in the diagram in terms of unit vectors **i** and **j**.

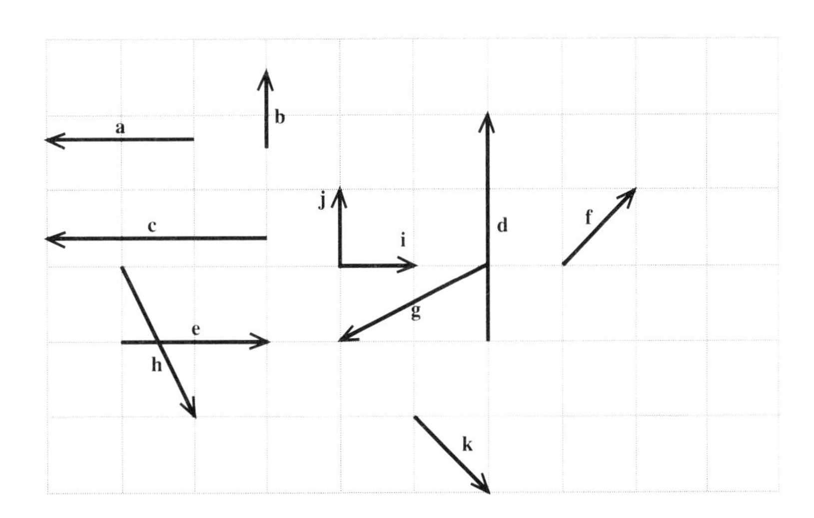

3. Given that $\mathbf{a} = \begin{pmatrix} 2 \\ -1 \end{pmatrix}$ and $\mathbf{b} = \begin{pmatrix} 1 \\ 4 \end{pmatrix}$ what are the coordinates of the point with position vector $3\mathbf{a} - 2\mathbf{b}$?

4. Four vectors are given in component form by

$\mathbf{a} = 3\mathbf{i} + 4\mathbf{j}$ $\quad \mathbf{b} = 6\mathbf{i} - 7\mathbf{j}$ $\quad \mathbf{c} = -2\mathbf{i} + 5\mathbf{j}$ $\quad \mathbf{d} = -5\mathbf{i} - 3\mathbf{j}$

Find the vectors:

(i) $\mathbf{a} + \mathbf{b}$, (ii) $\mathbf{b} + \mathbf{c}$, (iii) $\mathbf{c} + \mathbf{d}$,
(iv) $\mathbf{a} + \mathbf{b} + \mathbf{d}$, (v) $\mathbf{a} - \mathbf{b}$, (vi) $\mathbf{d} - \mathbf{b} + \mathbf{a}$.

5. Given vectors $\mathbf{a} = \begin{pmatrix} 4 \\ 1 \end{pmatrix}$, $\mathbf{b} = \begin{pmatrix} -1 \\ 0 \end{pmatrix}$, $\mathbf{c} = \begin{pmatrix} -2 \\ -3 \end{pmatrix}$ and $\mathbf{d} = \begin{pmatrix} 2 \\ 6 \end{pmatrix}$,

find

(i) $\mathbf{a} + 2\mathbf{b}$, (ii) $2\mathbf{c} - 3\mathbf{d}$,
(iii) $\mathbf{a} + \mathbf{c} - 2\mathbf{b}$, (iv) $-2\mathbf{a} + 3\mathbf{b} + 4\mathbf{d}$.

6. A, B and C are the points (1, 2), (5, 1) and (7, 8).
(i) Write down in terms of **i** and **j** the position vectors of these three points.
(ii) Find the component form of the displacements $\overrightarrow{AB}$, $\overrightarrow{BC}$ and $\overrightarrow{CA}$.
(iii) Show these six vectors on a diagram.

7. A, B and C are the points (0, −3), (2, 5), (3, 9).
(i) Write down in terms of **i** and **j** the position vectors of these three points.
(ii) Find the displacements $\overrightarrow{AB}$ and $\overrightarrow{BC}$.
(iii) Show that the three points all lie on a straight line.

Exercise 3B continued

8. Three vectors $\mathbf{a}$, $\mathbf{b}$ and $\mathbf{c}$ are given by $\mathbf{a} = \mathbf{i} + \mathbf{j}$, $\mathbf{b} = -\mathbf{i} + 2\mathbf{j}$ and $\mathbf{c} = 3\mathbf{i} - 4\mathbf{j}$. R is the end point of the displacement $2\mathbf{a} - 3\mathbf{b} + \mathbf{c}$ and (1, −2) is the starting point. What is the position vector of R?

9. Given the vectors $\mathbf{p} = 3\mathbf{i} - 5\mathbf{j}$ and $\mathbf{q} = -\mathbf{i} + 4\mathbf{j}$ find the vectors $\mathbf{x}$ and $\mathbf{y}$ where (i) $2\mathbf{x} - 3\mathbf{p} = \mathbf{q}$, (ii) $4\mathbf{p} - 3\mathbf{y} = 7\mathbf{q}$.

10. The vectors $\mathbf{x}$ and $\mathbf{y}$ are defined in terms of a and b as

$$\mathbf{x} = a\mathbf{i} + (a + b)\mathbf{j}$$

$$\mathbf{y} = (6 - b)\mathbf{i} - (2a + 3)\mathbf{j}$$

If $\mathbf{x} = \mathbf{y}$ find the values of a and b.

Resolution of vectors

We have seen how to represent vectors both in magnitude and direction form and in component form. How do you convert a vector from one form to the other?

Consider the vector $\mathbf{a}$ making an angle α with the x axis and β with the y axis as shown in figure 3.8. Note that α is the angle between the vector and the $\mathbf{i}$ direction and β is the angle between the vector and the $\mathbf{j}$ direction.

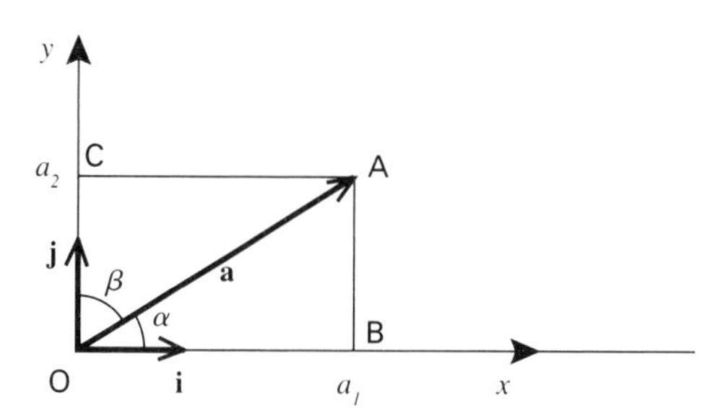

Figure 3.8

The vector $\mathbf{a}$ can be written as

$$\mathbf{a} = a_1\mathbf{i} + a_2\mathbf{j}$$

From triangle OAB, $a_1 = |\mathbf{a}|\cos\alpha = a\cos\alpha$.

From triangle OAC, $a_2 = |\mathbf{a}|\cos\beta = a\cos\beta$.

So the vector $\mathbf{a}$ can be written as

$$\mathbf{a} = a\cos\alpha\mathbf{i} + a\cos\beta\mathbf{j}.$$

The cosines of the angles α and β are called the *direction cosines* of the vector $\mathbf{a}$. Note that $\cos\beta = \sin\alpha$ so that $\mathbf{a}$ can also be written as

$$\mathbf{a} = a\cos\alpha\mathbf{i} + a\sin\alpha\mathbf{j}.$$

This process of splitting a vector into two components at right angles is called *resolving*, and is a very important technique.

The form $a\cos\alpha\mathbf{i} + b\sin\alpha\mathbf{j}$ is convenient when the angle β is not known, but the alternative $a\cos\alpha\mathbf{i} + a\cos\beta\mathbf{j}$ can be generalised into three dimensions, making it a powerful expression. Both forms are used in this book.

Converting from components into magnitude and direction

It is very easy to convert a vector from component form to magnitude and direction form.

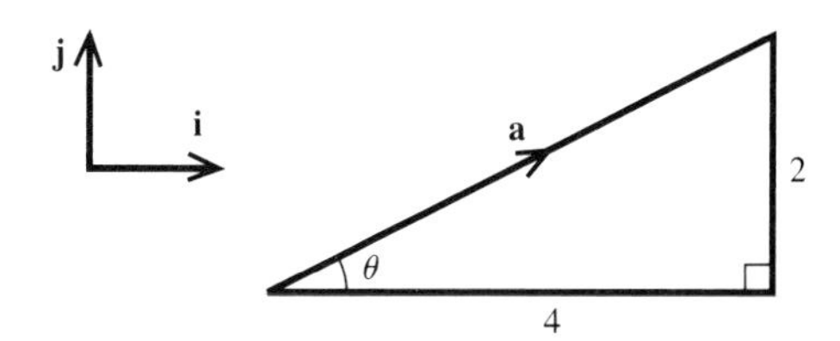

Figure 3.9

In figure 3.9, $\mathbf{a} = 4\mathbf{i} + 2\mathbf{j}$.

The magnitude of **a** is

$$|\mathbf{a}| = \sqrt{4^2 + 2^2} = \sqrt{20} \text{ (by Pythagoras' Theorem).}$$

The direction of **a** is given by

$$\tan\theta = \frac{2}{4} = \frac{1}{2}: \qquad \theta = 26.6°.$$

Using the notation in figure 3.10, this can be written in general form as the magnitude of the vector $a_1\mathbf{i} + a_2\mathbf{j} = \sqrt{{a_1}^2 + {a_2}^2}$;

the direction is given by $\theta = \arctan\left(\frac{a_2}{a_1}\right)$.

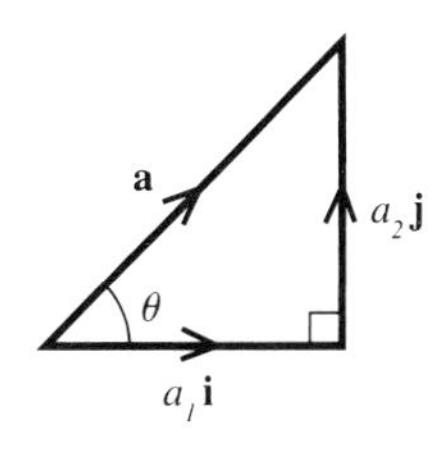

Figure 3.10

EXAMPLE

Two vectors **a** and **b** have magnitudes 4 and 5 in the directions shown in the diagram. Find the magnitude and direction of the vector $\mathbf{a}+\mathbf{b}$.

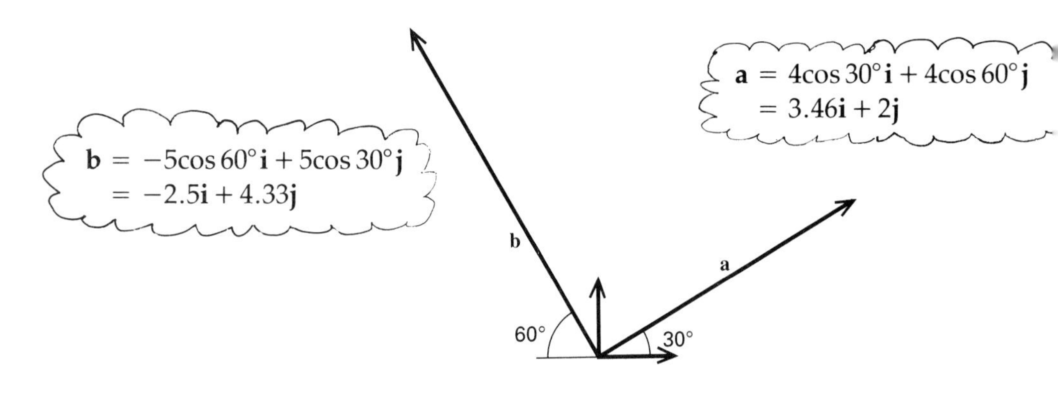

Solution

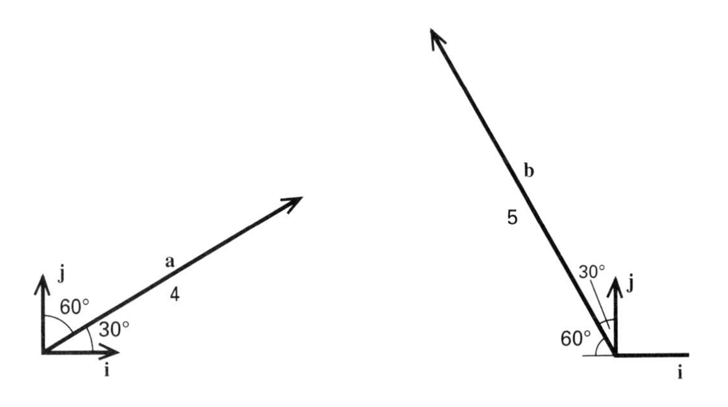

$$\begin{aligned}\mathbf{a}+\mathbf{b} &= (3.46\mathbf{i}+2\mathbf{j})+(-2.5\mathbf{i}+4.33\mathbf{j})\\ &= 0.96\mathbf{i}+6.33\mathbf{j}\end{aligned}$$

This vector is shown in the diagram on the right.

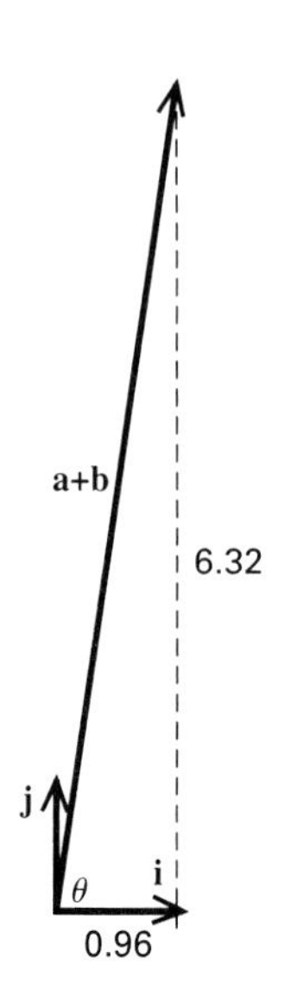

Magnitude $\quad |\mathbf{a}+\mathbf{b}| = \sqrt{0.96^2+6.33^2} = \sqrt{40.99} = 6.4$

Direction $\quad \tan\theta = \dfrac{6.33}{0.96} = 6.59$

$$\theta = 81.4°$$

The vector $\mathbf{a}+\mathbf{b}$ has magnitude 6.4 and direction 81.4° from the positive x direction.

NOTE *If the components in the **i** and **j** directions are not both positive, the angle the vector makes with the **i** direction is negative or greater than 90°*

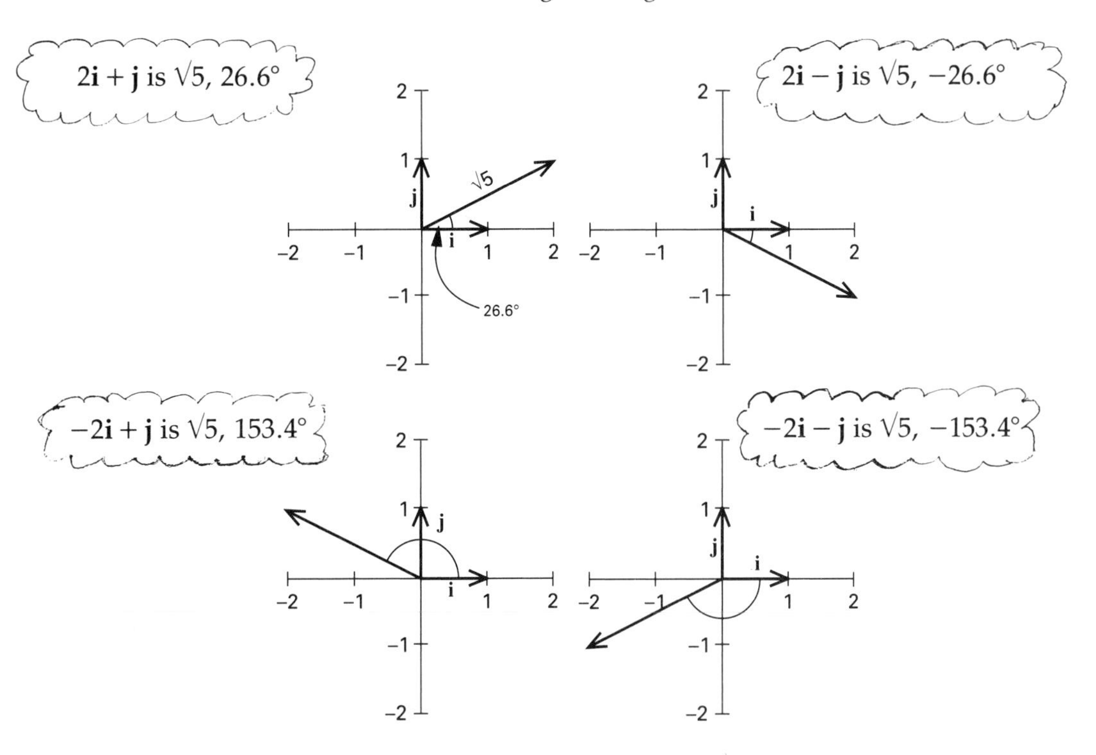

Vectors in three dimensions (space)

In three dimensions we choose the origin O of the xy plane and through it draw a third axis, perpendicular to both the x and y axes. It is conventional to draw the x, y and z axes as shown in figure 3.11A. The point Q (1, 3, 4) would then be plotted as shown in figure 3.11B.

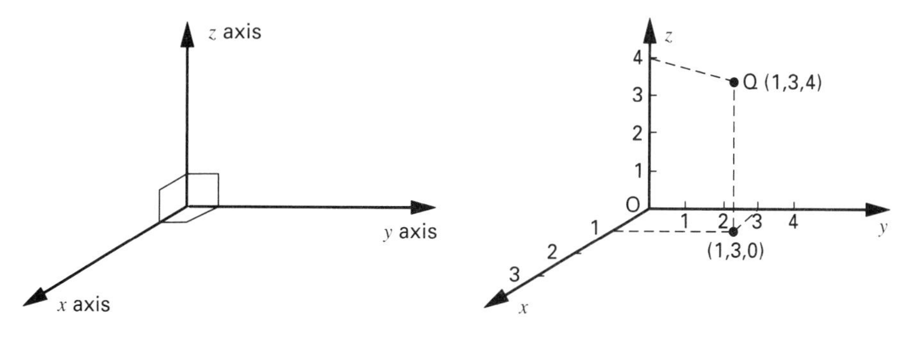

Figure 3.11A **Figure 3.11B**

A third unit vector **k** is introduced to represent the positive z direction. We now have three unit vectors **i**, **j** and **k** and any vector **a** can be written as

$$\mathbf{a} = a_1\mathbf{i} + a_2\mathbf{j} + a_3\mathbf{k}.$$

For example, the position vector of the point Q is written as

$$\overrightarrow{OQ} = \mathbf{i} + 3\mathbf{j} + 4\mathbf{k}.$$

In general the position vector of point P (x, y, z) is written as

$$\mathbf{r} = x\mathbf{i} + y\mathbf{j} + z\mathbf{k}.$$

EXAMPLE

The points A and B have position vectors $\mathbf{a} = 2\mathbf{i} - \mathbf{j} + 3\mathbf{k}$ and $\mathbf{b} = -2\mathbf{i} + \mathbf{k}$. Find $\overrightarrow{AB}$, $|\overrightarrow{OA}|$ and $|\overrightarrow{AB}|$

Solution

Using **i**, **j**, and **k**

$$\overrightarrow{AB} = \mathbf{b} - \mathbf{a}$$
$$= (-2\mathbf{i} + \mathbf{k}) - (2\mathbf{i} - \mathbf{j} + 3\mathbf{k})$$
$$= -4\mathbf{i} + \mathbf{j} - 2\mathbf{k}$$

Using column vectors

$$\overrightarrow{AB} = \mathbf{b} - \mathbf{a}$$
$$= \begin{pmatrix} -2 \\ 0 \\ 1 \end{pmatrix} - \begin{pmatrix} 2 \\ -1 \\ 3 \end{pmatrix}$$
$$= \begin{pmatrix} -4 \\ 1 \\ -2 \end{pmatrix}$$

$|\overrightarrow{OA}|$ is the magnitude of the vector $\overrightarrow{OA}$ and is found using Pythagoras' theorem on triangles OPQ and OPA shown in the diagram.

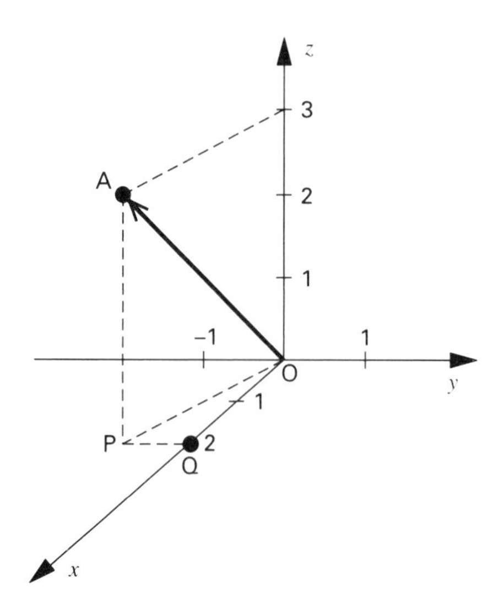

$$OP^2 = OQ^2 + QP^2 = 2^2 + (-1)^2 = 2^2 + 1^2$$
$$OA^2 = OP^2 + AP^2 = 2^2 + 1^2 + 3^2$$
$$\Rightarrow |OA| = \sqrt{14}$$

$|\overrightarrow{AB}|$ is the magnitude of vector $\overrightarrow{AB}$ and is also found using Pythagoras.

$$AB^2 = (-4)^2 + 1^2 + (-2)^2 = 4^2 + 1^2 + 2^2$$

$$\Rightarrow |AB| = \sqrt{21}$$

Note that in general if $\mathbf{a} = a_1\mathbf{i} + a_2\mathbf{j} + a_3\mathbf{k}$, then $|\mathbf{a}| = \sqrt{a_1^2 + a_2^2 + a_3^2}$

Unit vectors

Sometimes you need to write down a unit vector (i.e. a vector of magnitude 1) in the direction of a given vector. You have seen that the vector $a_1\mathbf{i} + a_2\mathbf{j}$ has magnitude $\sqrt{a_1^2 + a_2^2}$ and so it follows that the vector

$$\frac{a_1}{\sqrt{a_1^2 + a_2^2}}\mathbf{i} + \frac{a_2}{\sqrt{a_1^2 + a_2^2}}\mathbf{j}$$

has magnitude 1.

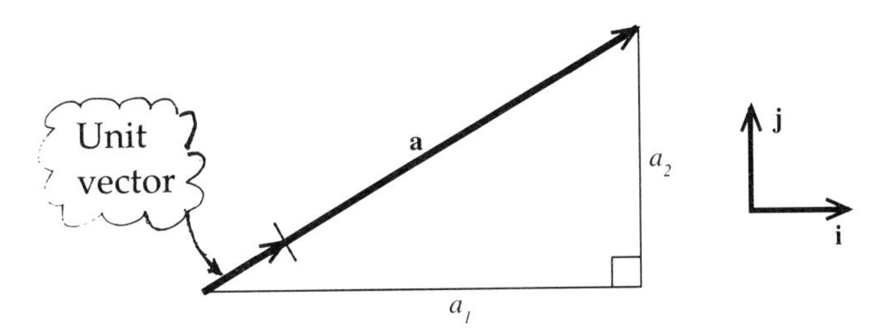

Similarly in 3 dimensions the unit vector in the direction of $a_1\mathbf{i} + a_2\mathbf{j} + a_3\mathbf{k}$ is

$$\frac{a_1}{\sqrt{a_1^2 + a_2^2 + a_3^2}}\mathbf{i} + \frac{a_2}{\sqrt{a_1^2 + a_2^2 + a_3^2}}\mathbf{j} + \frac{a_3}{\sqrt{a_1^2 + a_2^2 + a_3^2}}\mathbf{k}.$$

Exercise 3C

1. Using unit vectors **i** and **j** in directions east and north respectively, write the following displacements in component form.

(i) 130 km, bearing 060° (ii) 250 km, bearing 130°

(iii) 400 km, bearing 210° (iv) 50 miles, bearing 300°

Exercise 3C continued

2. Write down the following vectors in component form in terms of $\mathbf{i}$ and $\mathbf{j}$, and in column vector form.

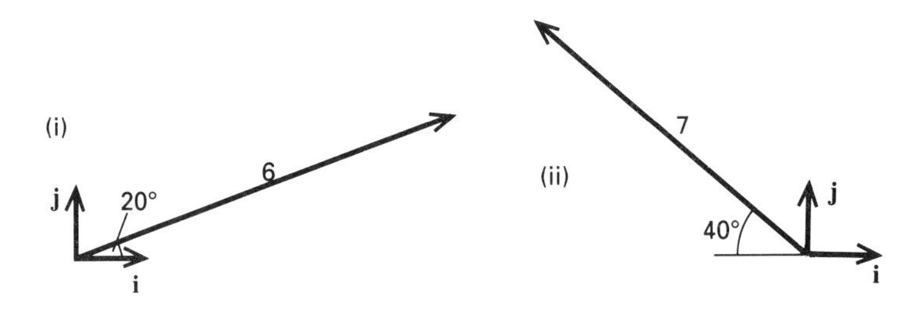

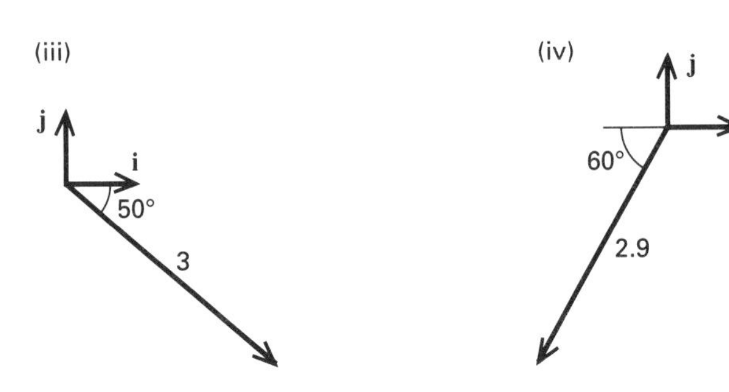

3. Write down each of the following vectors in terms of $\mathbf{i}$ and $\mathbf{j}$. Find the resultant of each system of vectors in terms of $\mathbf{i}$ and $\mathbf{j}$.

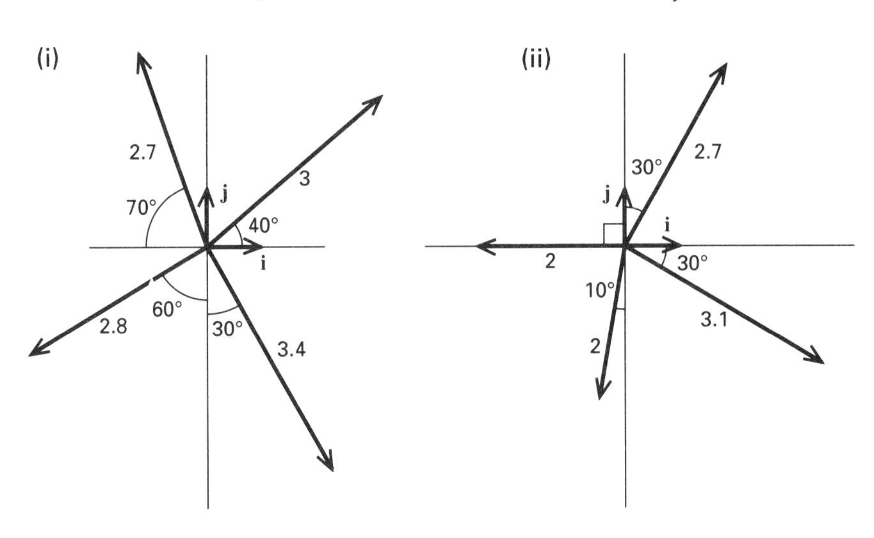

4. Calculate the magnitude and direction of the following vectors.

(i) $2\mathbf{i} + 3\mathbf{j}$ (ii) $4\mathbf{i} + 5\mathbf{j}$

(iii) $\mathbf{i} - \mathbf{j}$ (iv) $-2\mathbf{i} + 3\mathbf{j}$

(v) $\begin{pmatrix} 4 \\ -5 \end{pmatrix}$ (vi) $\begin{pmatrix} -2 \\ -6 \end{pmatrix}$

(vii) $\begin{pmatrix} -5 \\ 12 \end{pmatrix}$ (viii) $\begin{pmatrix} -3 \\ -4 \end{pmatrix}$

5. Find the distance and bearing of Sean relative to his starting point if he goes for a walk with the following three stages:
Stage 1 600 m on a bearing 30°,
Stage 2 1 km on a bearing 100°,
Stage 3 700 m on a bearing 340°.

Shona sets off from the same place at the same time as Sean, walking at the same speed, and takes the stages in the order 3–1–2. How far apart are Sean and Shona at the end of their walks?

6. The diagram shows the journey of a yacht.

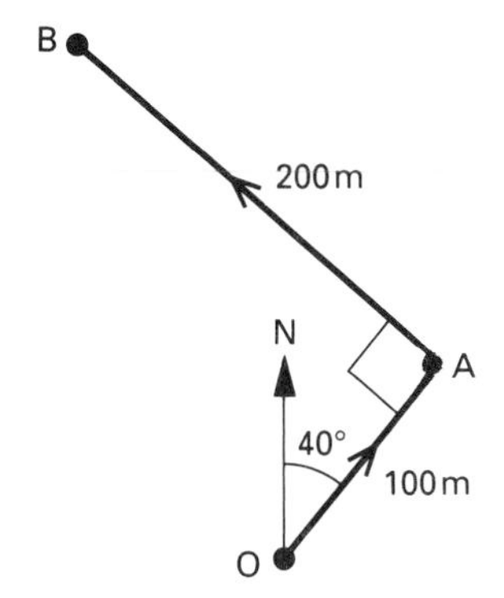

Express $\overrightarrow{OA}$, $\overrightarrow{AB}$ and $\overrightarrow{OB}$ as vectors in terms of **i** and **j**, which are unit vectors east and north respectively.

7. The diagram shows the top of a tower crane. Express the position vectors of A, B and C relative to O in terms of **i** and **j** as shown.

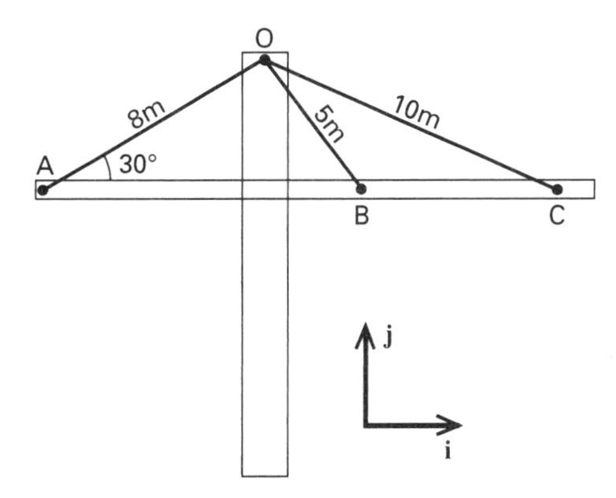

Exercise 3C continued

8. An aeroplane is travelling from Plymouth to London at 200 $\mathrm{km\,h^{-1}}$ on a bearing 075°. Due to fog the aeroplane changes direction to fly to Birmingham on a bearing of 015°.
   (i) Show the velocity vectors and the change in velocity on a diagram.
   (ii) Write down the components of the change in velocity in terms of unit vectors in directions east and north.

9. The diagram shows the big wheel ride from a fairground. The radius of the wheel is 5 m and the length of the arms that support each chair is 1 m. Express the position vector of the carriages indicated in terms of **i** and **j**.

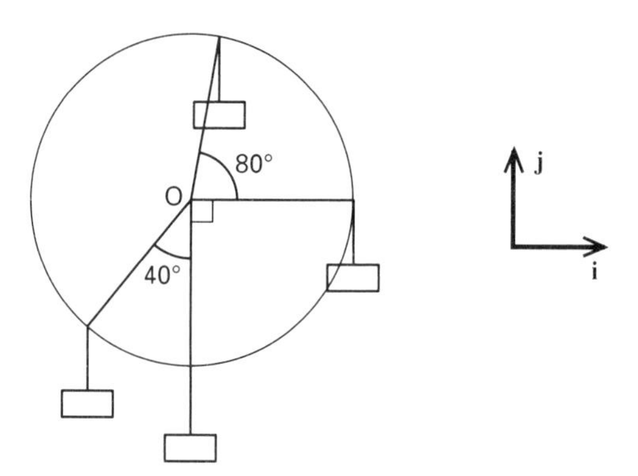

10. A yacht sails 50 m north-east and then 100 m on a bearing of 300°.
   (i) Express each displacement in terms of unit vectors **i** (east) and **j** (north).
   (ii) Express the total displacement in terms of **i** and **j**.
   (iii) How far, and in what direction, would the yacht have sailed if it had travelled directly from its starting point to its final position?

11. An aeroplane completes a journey in three stages. The displacements at each stage, in kilometres, are $3000\mathbf{i} + 4000\mathbf{j}$, $1000\mathbf{i} + 500\mathbf{j}$ and $300\mathbf{i} - 1000\mathbf{j}$, where **i** and **j** are unit vectors in directions east and north respectively.

Express the total journey as a distance and a bearing.

12. Two walkers set off from the same place in different directions. After a period they stop. Their displacements are $\begin{pmatrix} 2 \\ 5 \end{pmatrix}$ and $\begin{pmatrix} -3 \\ 4 \end{pmatrix}$ where the distances are in kilometres and the directions are east and north.

On what bearing and for what distance does the second walker have to walk to be reunited with the first (who does not move)?

13. The position vectors of the points A, B and C are $\mathbf{a} = \mathbf{i} + \mathbf{j} - 2\mathbf{k}$, $\mathbf{b} = 6\mathbf{i} - 3\mathbf{j} + \mathbf{k}$ and $\mathbf{c} = -2\mathbf{i} + 2\mathbf{j}$.
Find
(i) the vectors $\overrightarrow{AC}$, $\overrightarrow{AB}$ and $\overrightarrow{BC}$,
(ii) $|\mathbf{a}|$, $|\mathbf{b}|$ and $|\mathbf{c}|$.
(iii) Show that $|\mathbf{a} + \mathbf{b} - \mathbf{c}|$ is **not** equal to $|\mathbf{a}| + |\mathbf{b}| - |\mathbf{c}|$.

14. (i) Show that $\frac{3}{5}\mathbf{i} + \frac{4}{5}\mathbf{j}$ is a unit vector.

(ii) Find unit vectors in the directions of
(a) $8\mathbf{i} + 6\mathbf{j}$ (b) $5\mathbf{i} - 12\mathbf{j}$

15. (i) Show that $\frac{1}{3}\mathbf{i} - \frac{2}{3}\mathbf{j} + \frac{2}{3}\mathbf{k}$ is a unit vector.

(ii) Find unit vectors in the directions of
(a) $2\mathbf{i} - 6\mathbf{j} + 7\mathbf{k}$ (b) $\mathbf{i} + \mathbf{j} + \mathbf{k}$

Resultant velocity

Have you ever tried canoeing across a river with a strong current? It is quite difficult to canoe directly across to a point on the other bank of the river. To calculate your actual path you need to find your *resultant velocity*.

EXAMPLE

A swimmer is attempting to cross a river which has a current of 5 kmh^{-1} parallel to its banks. She aims at a point directly opposite her starting point and can swim at 4 kmh^{-1} in still water. Find the resultant velocity of the swimmer, in the form of a speed and a bearing.

Solution

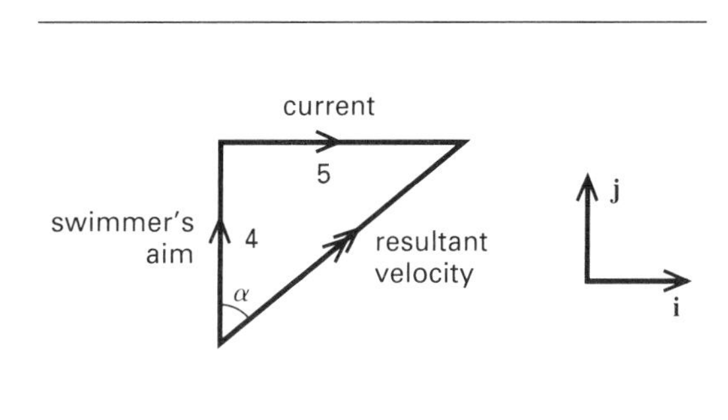

Choose unit vectors **i** and **j** parallel and perpendicular to the river bank. Then in component form

$$\text{velocity of current} = 5\mathbf{i}$$

$$\text{velocity of swimmer} = 4\mathbf{j}$$

$$\Rightarrow \quad \text{resultant velocity} = 5\mathbf{i} + 4\mathbf{j}.$$

$$\text{Actual speed} = \sqrt{5^2 + 4^2}$$

$$\approx 6.4\ \text{kmh}^{-1}$$

$$\text{Direction:} \quad \tan\alpha = \frac{5}{4} \text{ so } \alpha \approx 51°$$

The swimmer has speed 6.4 kmh^{-1} at an angle of 39° to the bank.

EXAMPLE

A small motor boat moving at 8 kmh^{-1} relative to the water travels directly between two lighthouses which are 10 km apart, the bearing of the second lighthouse from the first being 135°. The current has a constant speed of 4 kmh^{-1} from the east. Find (i) the course that the boat must set, and (ii) the time for the journey.

Solution

It is helpful to start a question like this with an accurate drawing. The boat is required to achieve a course of 135° and this must be the direction of the resultant of the boat's own velocity and that of the current.

First draw a line representing this resultant and mark on it a point A. Then draw a line AB, where $\overrightarrow{AB}$ represents the current (4 kmh^{-1} from the east). Finally, draw a line BC, where $\overrightarrow{BC}$ represents the boat's velocity (8 kmh^{-1}); this must be to the same scale as AB, and C must lie on the resultant. (If you were to solve the problem by scale drawing you would need to use a compass to find point C.)

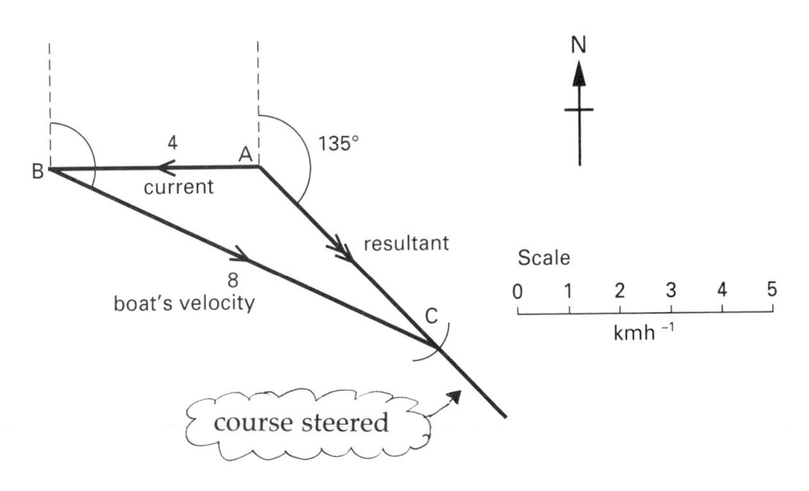

(i) To calculate the course, use the Sine Rule in triangle ABC: $\angle BAC = 135°$, $AB = 4$, $BC = 8$.

$$\frac{\sin \angle ACB}{4} = \frac{\sin 135°}{8} \Rightarrow \angle ACB = 20.7°$$

$\angle ABC = 180° - 135° - 20.7° = 24.3°$

Therefore the course steered $= 90° + 24° = 114°$ (to the nearest degree).

(ii) The resultant speed, AC, is also found by the Sine Rule.

$$\frac{AC}{\sin 24.3°} = \frac{8}{\sin 135°}$$

$AC = 4.656 \text{ kmh}^{-1}$

$$\text{Journey time} = \frac{10}{4.656} \text{ h} = 2.15 \text{ hours.}$$

Exercise 3D

1. A yacht is sailing at a speed of 6 knots due east, but the current is flowing north-west at a speed of 3 knots. Find the resultant speed of the yacht.

2. An aeroplane flies due north at 300 kmh^{-1}, but a crosswind blows north-west at 40 kmh^{-1}. Find the resultant velocity of the aeroplane.

Exercise 3D continued

3. A child runs up and down a train. If the child runs at 2 ms^{-1} and the train moves at 30 ms^{-1}, what are the resultant velocities of the child?

4. A machine-gun fires bullets at a speed of 100 ms^{-1}, from an aeroplane that is moving at 50 ms^{-1}. The bullets can be fired in any direction: what range of speeds do they have?

5. A ship has speed 10 knots in still water. It heads due west in a current of 3 knots from 150°.
(i) Find the actual speed and course of the ship.
(ii) If the ship is to travel due west, on what course must the captain steer?

6. A man can paddle a canoe at 3 ms^{-1} in still water. He wishes to paddle straight across a river which is 20 m wide and flows at a constant speed of 1.5 ms^{-1} parallel to the banks.
(i) At what angle to the bank should he paddle?
(ii) How long does he take to cross?

7. A small motorboat moving at 8 kmh^{-1} relative to the water travels directly between two lighthouses which are 10 km apart, the bearing of the second lighthouse from the first being 150°. The current has a constant speed of 4 kmh^{-1} from the west. Find the course that the boat must set and the time for the journey.

8. An aeroplane leaves Heathrow airport at noon travelling due east towards Prague, 1200 km away. The speed of the aeroplane in still air is 400 kmh^{-1}. Initially there is no wind.
(i) Estimate the time of arrival of the plane.

After flying for one hour the aeroplane runs into a storm with strong winds of 75 kmh^{-1} from the south-west.
(ii) If the pilot fails to adjust the heading of the aeroplane how many kilometres from Prague will the plane be after the estimated journey time?
(iii) What new course should the pilot set?

9. A car ferry travels between ports A and B on two islands, which are 60 km apart. Port A is due west of port B. There is a current in the sea of constant speed 4 kmh^{-1} from a bearing 260°. The car ferry can travel at 20 kmh^{-1} in still water.
(i) Calculate the bearing on which the ferry must head from each port.
(ii) On a particular journey from A to B, the engines fail when the ferry is half way between the ports. If the coastline near to port B runs north–south, at what distance from port B does the ferry hit land?

Exercise 3D continued

10. Rain falling vertically hits the windows of an Intercity train which is travelling at 50 ms^{-1}. A passenger watches the rain streaks on the window and estimates them to be at an angle of 20° to the horizontal.

(i) Calculate the speed of the rain.

(ii) The train starts to slow down with a uniform deceleration of 0.3 ms^{-2}. Find an expression for the angle of the rain streaks to the horizontal t seconds later.

KEY POINTS

- A vector has both magnitude and direction.
- Vectors may be represented in magnitude–direction form or in component form.

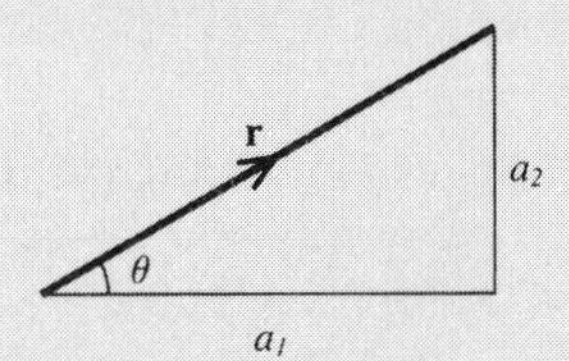

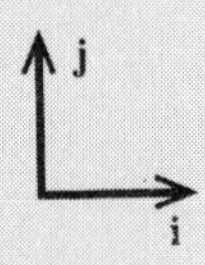

Magnitude–direction form

Magnitude r direction θ

where $r = \sqrt{a_1^2 + a_2^2}$

$\tan\theta = \dfrac{a_2}{a_1}$

Component form

$\begin{pmatrix} a_1 \\ a_2 \end{pmatrix}$ or $a_1\mathbf{i} + a_2\mathbf{j}$

where $a_1 = r\cos\theta$

$a_2 = r\sin\theta$

$= r\cos(90° - \theta)$

- When two or more vectors are added the **resultant** is obtained. Vector addition may be done graphically or algebraically.

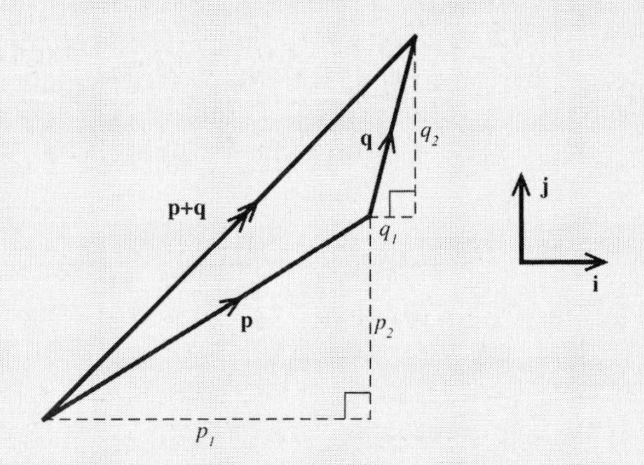

$$\mathbf{p}+\mathbf{q} = \begin{pmatrix} p_1 \\ p_2 \end{pmatrix} + \begin{pmatrix} q_1 \\ q_2 \end{pmatrix} = \begin{pmatrix} p_1+q_1 \\ p_2+q_2 \end{pmatrix}$$

$$\text{or } \mathbf{p}+\mathbf{q} = (p_1+q_1)\mathbf{i} + (p_2+q_2)\mathbf{j}$$

- The position vector is the vector representing the displacement of a point from the origin.
- Equal vectors have the same magnitude and direction and are equal component by component.
- The unit vector in the direction of $a_1\mathbf{i} + a_2\mathbf{j}$ is given by

$$\frac{a_1}{\sqrt{a_1{}^2 + a_2{}^2}}\,\mathbf{i} + \frac{a_2}{\sqrt{a_1{}^2 + a_2{}^2}}\,\mathbf{j}$$

4 Motion in two and three dimensions

There was a young lady named Bright
Whose speed was far faster than light,
She set out one day in a relative way
And returned home the previous night.

A. H. R. Buller – on motion in four dimensions

Think of three physical situations that involve the motion of an object in two dimensions. Sketch the path of the object.

What can you say about the position, velocity and acceleration of the object at any time? What questions might you need to answer about the object's motion?

In Chapter 2 you explored the motion of an object moving along a straight line. Many real problems in mechanics involve the motion of objects in two or three dimensions.

As an example consider the traffic moving on the roundabout in the photograph. It is fairly obvious that the vehicles are not travelling in straight lines. As there is very little vertical movement their motion could be modelled as motion in a horizontal plane, i.e. motion in two dimensions.

Velocity

velocity $15\cos 60°\mathbf{i} + 15\cos 30°\mathbf{j} = 7.5\mathbf{i} + 12.99\mathbf{j}$

$\sqrt{7.5^2 + 12.99^2} = 15$

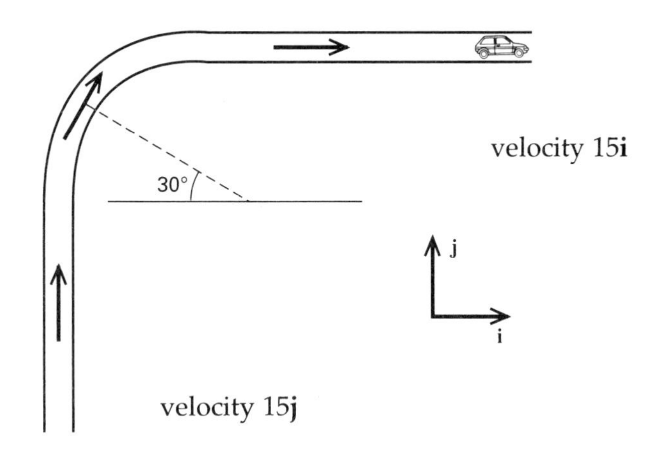

Figure 4.1

A car travels along a straight stretch of road, goes round a bend and then leaves on another straight stretch, at a constant speed of 15 ms^{-1} throughout. The magnitude of the velocity is constant but its direction changes, always directed along the tangent to the curve as shown in figure 4.1. Once the car has rounded the bend its velocity is again constant. The velocity vectors are shown at three positions of the car; they all have the same magnitude because the car's speed is constant, but different directions.

Velocity is a vector in the direction of the motion with magnitude equal to the speed.

For Discussion

1. Look at this sketch of the path of a ball. At the points marked, in what directions would you draw arrows to represent the velocity of the ball?

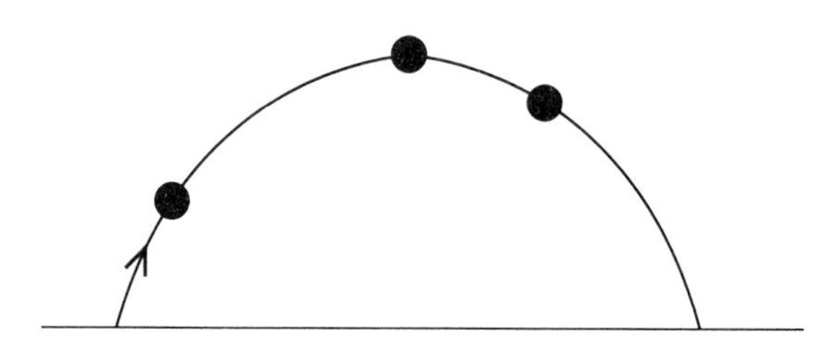

For Discussion continued

2. Describe what happens to the speed and velocity of an athlete:
(i) sprinting in a 100 m race;
(ii) running in a 400 m race;
(iii) running a marathon.

The velocity vector

In one dimension velocity is calculated as rate of change of position. The same is true in two dimensions where we split vectors into two components using unit vectors **i** and **j**, so that the position vector is $\mathbf{r} = x\mathbf{i} + y\mathbf{j}$.

In two dimensions,

$$\text{velocity } \mathbf{v} = \frac{\mathrm{d}\mathbf{r}}{\mathrm{d}t} = \frac{\mathrm{d}x}{\mathrm{d}t}\mathbf{i} + \frac{\mathrm{d}y}{\mathrm{d}t}\mathbf{j}$$

Similarly in three dimensions,

$$\text{velocity } \mathbf{v} = \frac{\mathrm{d}\mathbf{r}}{\mathrm{d}t} = \frac{\mathrm{d}x}{\mathrm{d}t}\mathbf{i} + \frac{\mathrm{d}y}{\mathrm{d}t}\mathbf{j} + \frac{\mathrm{d}z}{\mathrm{d}t}\mathbf{k}$$

EXAMPLE

The position (in metres) of a ball is modelled by

$$\mathbf{r} = 5t\mathbf{i} + (12t - 5t^2)\mathbf{j}$$

where t is the time in seconds.
(i) Find an expression for the velocity of the ball at time t.
(ii) Calculate the velocity and speed of the ball when $t = 0$, 1 and 2.
(iii) Draw a sketch of the path of the ball and include vectors to illustrate the position and the direction of the velocity of the ball at time $t = 2$.
(iv) Find the equation of the trajectory of the ball.

Solution

(i) To obtain the velocity vector we differentiate the components of the position vector to give

$$\mathbf{v} = \frac{\mathrm{d}\mathbf{r}}{\mathrm{d}t} = 5\mathbf{i} + (12 - 10t)\mathbf{j}.$$

(ii) Substituting for $t = 0$, 1 and 2 gives

$t = 0$: velocity $\mathbf{v} = 5\mathbf{i} + 12\mathbf{j}$ with speed $|\mathbf{v}| = \sqrt{5^2 + 12^2} = 13\ \text{ms}^{-1}$

$t = 1$: velocity $\mathbf{v} = 5\mathbf{i} + 2\mathbf{j}$ with speed $|\mathbf{v}| = \sqrt{5^2 + 2^2} = \sqrt{29}\ \text{ms}^{-1}$

$t = 2$: velocity $\mathbf{v} = 5\mathbf{i} - 8\mathbf{j}$ with speed $|\mathbf{v}| = \sqrt{5^2 + (-8)^2} = \sqrt{89}\ \text{ms}^{-1}$.

(iii) The position and velocity vectors at time $t = 2$ are shown on the sketch of the path of the ball.

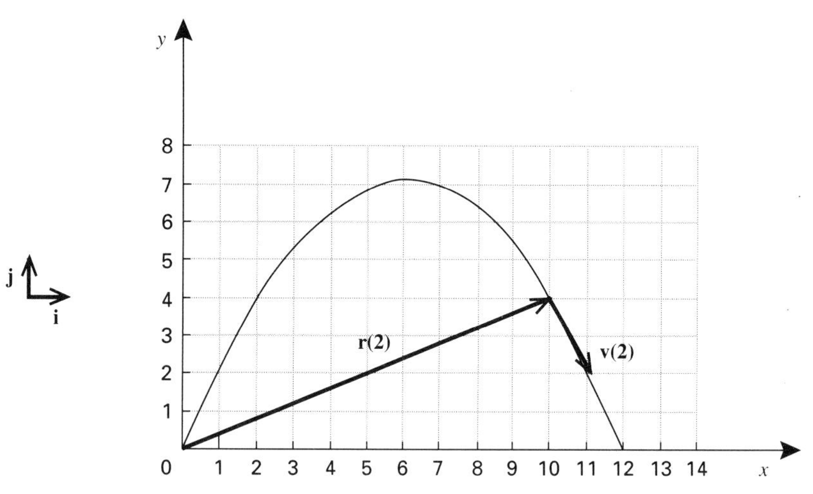

(iv) To find the equation of the trajectory, use

$$\mathbf{r} = x\mathbf{i} + y\mathbf{j} = 5t\mathbf{i} + (12t - 5t^2)\mathbf{j}$$

to get $x = 5t$ and $y = 12t - 5t^2$.

Since $x = 5t$, $t = \dfrac{x}{5}$ and so $y = 12\left(\dfrac{x}{5}\right) - 5\left(\dfrac{x}{5}\right)^2 = \dfrac{1}{5}(12x - x^2)$

Acceleration

The acceleration of an object is also a vector quantity with magnitude and direction. Again the one-dimensional definition of acceleration as the rate of change of velocity extends to two and three dimensions. However the direction of the acceleration vector is often not obvious. Consider the motion of an object travelling along a curved path so that at points P and Q the velocity vectors are $\mathbf{v_P}$ and $\mathbf{v_Q}$ respectively, as shown in figure 4.2. The change in velocity between P and Q is the vector $\mathbf{v_Q} - \mathbf{v_P}$. This is shown in figure 4.3.

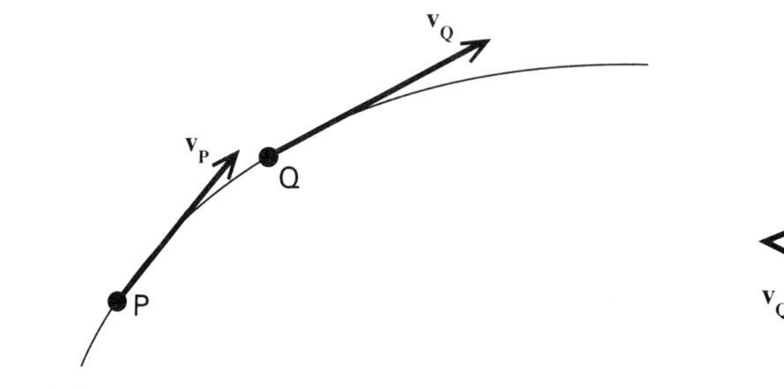

Figure 4.2

Figure 4.3

If this change in velocity takes place over t seconds the average acceleration is $(\mathbf{v_P} - \mathbf{v_Q})/t$. The direction of this vector is **not** along the path of the object. The actual direction depends on the shape of the curve along which the object is moving. For example, in the special case of circular motion with constant speed the direction of the acceleration is towards the centre of the circle, see figure 4.4.

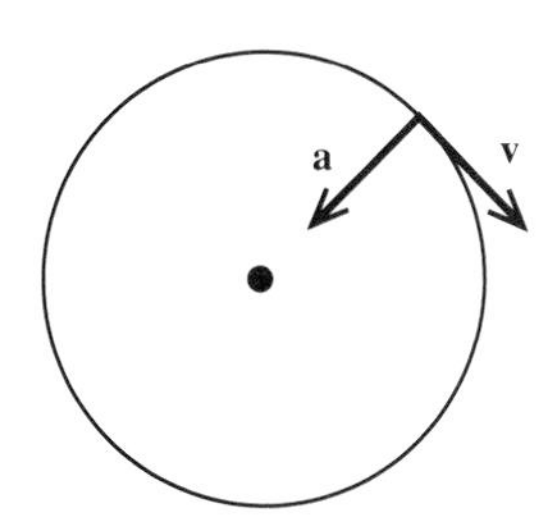

Figure 4.4

EXAMPLE

A car travels round a roundabout so that its speed increases from 5 ms^{-1} to 12 ms^{-1} as it travels one quarter of a complete circle in 4 seconds. Find the average acceleration of the car during this period.

Solution

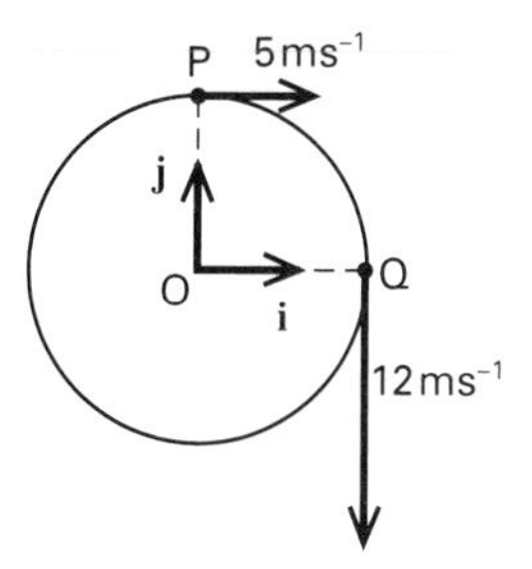

The average acceleration is the vector $(\mathbf{v_Q} - \mathbf{v_P})/4$. Choosing unit vectors $\mathbf{i}$ and $\mathbf{j}$ as shown, $\mathbf{v_P} = 5\mathbf{i}$ and $\mathbf{v_Q} = -12\mathbf{j}$ so that

$$\text{average acceleration} = \frac{\mathbf{v_Q} - \mathbf{v_P}}{4} = \frac{-12\mathbf{j} - 5\mathbf{i}}{4}.$$

This vector is shown in the diagram.

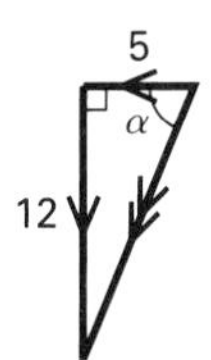

$$\text{Magnitude} = \sqrt{\left(\frac{5}{4}\right)^2 + \left(\frac{12}{4}\right)^2} = 3.25 \text{ ms}^{-2}$$

$$\text{Direction: } \tan\alpha = \frac{12}{3}$$

$$\alpha = 67° \text{ (to the nearest degree).}$$

Compass bearing = 203°.

As the car goes round the roundabout its acceleration and velocity are constantly changing in both magnitude and direction. You have seen a method for finding the average acceleration over a finite period, but how do you find the instantaneous acceleration?

The car's velocity is a function of time and its acceleration is the rate of change of its velocity. This means that, as in the one-dimensional case, an expression for the acceleration may be found by differentiating the velocity with respect to time. To find the acceleration at a particular instant (say $t = 3$) you then substitute $t = 3$ into the expression.

If **v** and **a** are the velocity and acceleration vectors, then

$$\mathbf{a} = \frac{\mathrm{d}\mathbf{v}}{\mathrm{d}t}$$

EXAMPLE

An aircraft is dropping a crate of supplies onto level ground. Relative to an observer on the ground, the crate is released at the point with position vector $\begin{pmatrix} 650 \\ 576 \end{pmatrix}$ m with initial velocity $\begin{pmatrix} -100 \\ 0 \end{pmatrix}$ ms^{-1}, where the directions are horizontal and vertical. Its acceleration is modelled by

$$\mathbf{a} = \begin{pmatrix} -t + 12 \\ \frac{1}{2}t - 10 \end{pmatrix} \text{ for } t \leqslant 12 \text{ seconds.}$$

(i) Find an expression for the velocity vector of the crate at time t.
(ii) Find an expression for the position vector of the crate at time t.
(iii) Show that the crate hits the ground 12 seconds after its release and find how far from the observer this happens.

Solution

(i) $\mathbf{a} = \dfrac{\mathrm{d}\mathbf{v}}{\mathrm{d}t} = \begin{pmatrix} -t + 12 \\ \frac{1}{2}t - 10 \end{pmatrix}$

Integrating gives $\mathbf{v} = \begin{pmatrix} -\frac{1}{2}t^2 + 12t + c_1 \\ \frac{1}{4}t^2 - 10t + c_2 \end{pmatrix}$

At $t = 0$, $\mathbf{v} = \begin{pmatrix} -100 \\ 0 \end{pmatrix} \Rightarrow \begin{matrix} c_1 = -100 \\ c_2 = 0 \end{matrix}$

$\therefore$ Velocity, $\mathbf{v} = \begin{pmatrix} -\frac{1}{2}t^2 + 12t - 100 \\ \frac{1}{4}t^2 - 10t \end{pmatrix}$

(ii) Since $\mathbf{v} = \dfrac{\mathrm{d}\mathbf{r}}{\mathrm{d}t}$, integrating again gives

$$\mathbf{r} = \begin{pmatrix} -\frac{1}{6}t^3 + 6t^2 - 100t + k_1 \\ \frac{1}{12}t^3 - 5t^2 + k_2 \end{pmatrix}$$

At $t = 0$, $\mathbf{r} = \begin{pmatrix} 650 \\ 576 \end{pmatrix} \Rightarrow \begin{matrix} k_1 = 650 \\ k_2 = 576 \end{matrix}$

$\therefore$ Position vector $\mathbf{r} = \begin{pmatrix} -\frac{1}{6}t^3 + 6t^2 - 100t + 650 \\ \frac{1}{12}t^3 - 5t^2 + 576 \end{pmatrix}$

(iii) At $t = 12$, $\mathbf{r} = \begin{pmatrix} -\frac{1}{6} \times 12^3 + 6 \times 12^2 - 100 \times 12 + 650 \\ \frac{1}{12} \times 12^3 - 5 \times 12^2 + 576 \end{pmatrix} = \begin{pmatrix} 26 \\ 0 \end{pmatrix}$

$\therefore$ The crate hits the ground at time $t = 12$, 26 m in front of the observer.

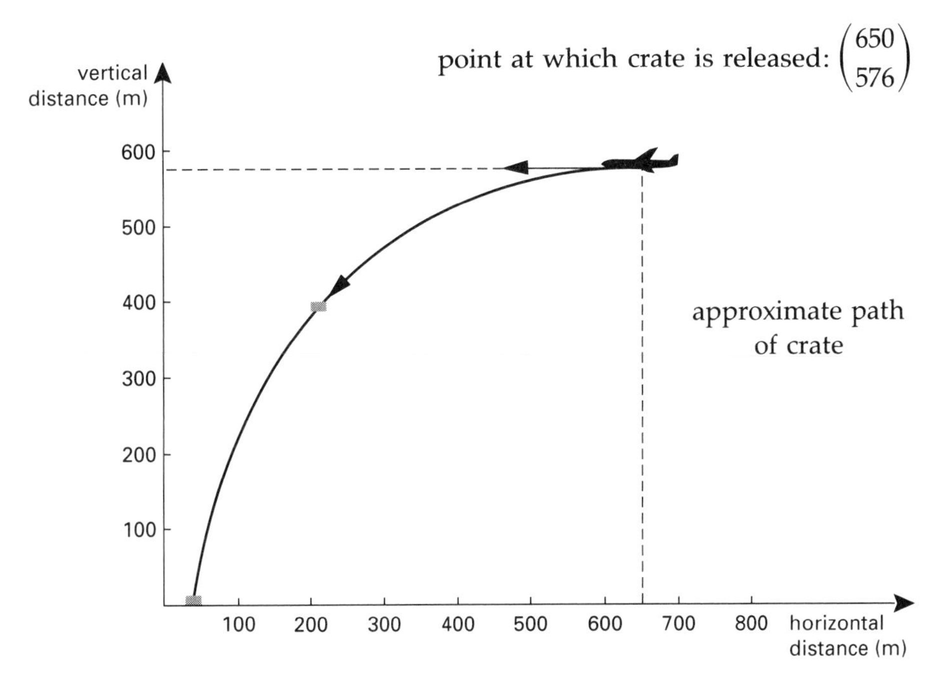

point at which crate hits ground: $\begin{pmatrix} 26 \\ 0 \end{pmatrix}$

Vector equations of constant acceleration

If the acceleration $\mathbf{a}$ is constant then the equations of constant acceleration can be written in vector notation as follows.

$$\mathbf{v} = \mathbf{u} + \mathbf{a}t$$

$$\mathbf{r} = \mathbf{u}t + \tfrac{1}{2}\mathbf{a}t^2$$

$$\mathbf{r} = \tfrac{1}{2}(\mathbf{u} + \mathbf{v})t\,.$$

You will find these forms useful in Chapter 8 on projectiles.

Exercise 4A

1. Write the following vectors in component form choosing unit vectors **i** and **j** in directions east and north.

Exercise 4A continued

(i) velocity 40 kmh^{-1} on a bearing 030°
(ii) velocity 12 ms^{-1} on a bearing 270°
(iii) acceleration 6 ms^{-2} on a bearing 130°
(iv) acceleration 10 ms^{-2} due south.

2. A shuttlecock is hit into the air with initial speed 10 ms^{-1} and at an angle of 40° to the horizontal. Express its initial velocity in terms of $\mathbf{i}$ and $\mathbf{j}$, unit vectors that are horizontal and vertical. It hits the floor with velocity $3\mathbf{i} - 4\mathbf{j}$ (in metres per second). Find its speed and direction when it hits the floor.

3. The first part of a race track is a bend. As the leading car travels round the bend its position, in metres, is modelled by:

$$\mathbf{r} = 2t^2\mathbf{i} + 8t\mathbf{j}$$

where t is in seconds.

(i) Find an expression for the velocity of the car.
(ii) Find the position of the car when $t = 0, 1, 2, 3$ and 4. Use this information to sketch the path of the car.
(iii) Find the velocity of the car when $t = 0, 1, 2, 3$ and 4. Add vectors to your sketch to represent these velocities.
(iv) Find the speed of the car as it leaves the bend at $t = 5$.

4. Find the average acceleration between points P and Q for an object travelling on each of the following circular paths, if it takes 3 seconds to travel between P and Q in each case.

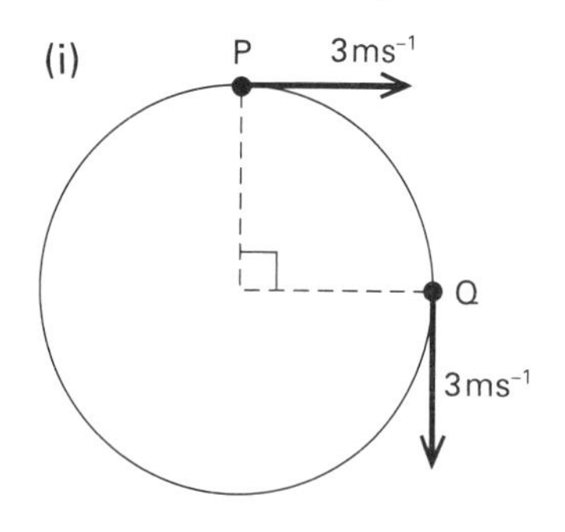

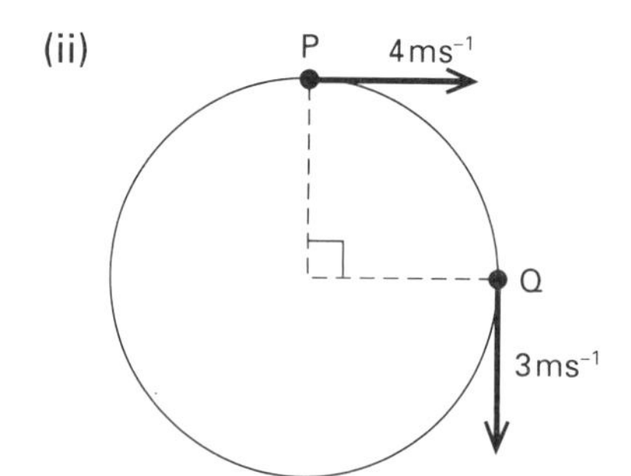

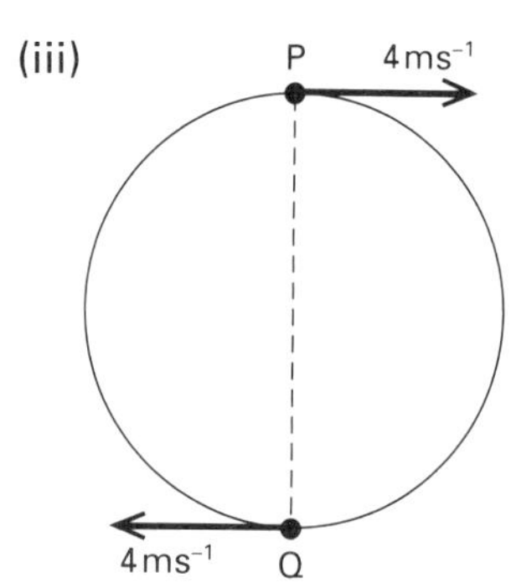

5. As a boy slides down a slide his position vector in metres at time t is

$$\begin{pmatrix} x \\ y \end{pmatrix} = \begin{pmatrix} 16 - 4t \\ 20 - 5t \end{pmatrix}$$

Find his velocity and acceleration.

6. Calculate the magnitude and direction of the acceleration of a particle that moves so that its position vector in metres is given by

$$\mathbf{r} = (8t - 2t^2)\mathbf{i} + (6 + 4t - t^2)\mathbf{j}$$

where t is the time in seconds.

7. A rocket moves with a velocity (in ms^{-1}) modelled by

$$\mathbf{v} = \frac{1}{10}t\mathbf{i} + \frac{1}{10}t^2\mathbf{j}$$

where $\mathbf{i}$ and $\mathbf{j}$ are horizontal and vertical unit vectors respectively, and t is in seconds. Find

(i) an expression for its position vector relative to its starting position at time t,

(ii) the displacement of the rocket after ten seconds of its flight.

8. A particle is initially at rest at the origin. It experiences an acceleration given by

$$\mathbf{a} = 4t\mathbf{i} + (6 - 2t)\mathbf{j}.$$

Find expressions for the velocity and position of the particle at time t.

9. While a hockey ball is being hit it experiences an acceleration (in ms^{-2}) modelled by

$\mathbf{a} = 1000[6t(t - 0.2)\mathbf{i} + t(t - 0.2)\mathbf{j}]$ for $0 \leqslant t \leqslant 0.2$ in seconds.

If the ball is initially at rest, find the speed of the ball when it loses contact with the stick after 0.2 seconds.

10. A speedboat is initially moving at $5\ \text{ms}^{-1}$ on a bearing of 135°. Express the initial velocity as a vector in terms of $\mathbf{i}$ and $\mathbf{j}$, which are unit vectors east and north respectively. The boat then begins to accelerate with an acceleration modelled by

$$\mathbf{a} = \frac{1}{10}t\mathbf{i} + \frac{3}{10}t\mathbf{j} \quad \text{in ms}^{-2}.$$

Find the velocity of the boat 10 seconds after it begins to accelerate and its displacement over the 10 second period.

11. The layout of the British motor racing Grand Prix track at Silverstone is shown in the diagram. Make a copy of this diagram and show by an arrow the direction of the acceleration and velocity vectors at each of the marked points. (Remember that racing cars will slow down as they enter a corner and accelerate as they leave a corner.)

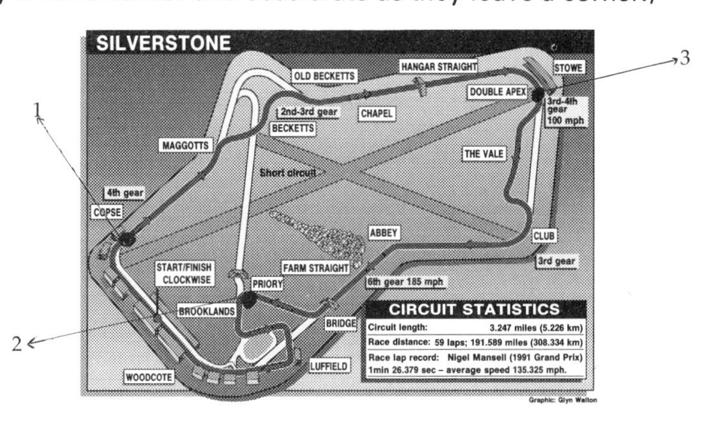

Exercise 4A continued

12. A girl throws a ball, and t seconds after she releases it its position in metres relative to the point where she is standing is modelled by

$$\begin{pmatrix} x \\ y \end{pmatrix} = \begin{pmatrix} 15t \\ 2 + 16t - 5t^2 \end{pmatrix}$$

where the directions are horizontal and vertical.

(i) Find expressions for the velocity and acceleration of the ball at time t.

(ii) The vertical component of the velocity is zero when the ball is at its highest point. Find the time taken for the ball to reach this point.

(iii) When the ball hits the ground the vertical component of its position vector is zero. What is the speed of the ball when it hits the ground?

(iv) Find the equation of the trajectory of the ball.

13. The position (in metres) of a tennis ball t seconds after leaving the racquet is modelled by

$$\mathbf{r} = (20t)\mathbf{i} + (2 + t - 5t^2)\mathbf{j}$$

where $\mathbf{i}$ and $\mathbf{j}$ are horizontal and vertical unit vectors.

(i) Find the position of the tennis ball when $t = 0, 0.2, 0.4, 0.6$ and 0.8. Use these to sketch the path of the ball.

(ii) Find an expression for the velocity of the tennis ball. Use this to find the velocity of the ball when $t = 0.2$.

(iii) Find the acceleration of the ball.

(iv) Find the equation of the trajectory of the ball.

14. John hits a golf ball over level ground with initial velocity $40\mathbf{i} + 30\mathbf{k}$ ms^{-1}, where the vector $\mathbf{i}$ is horizontal in the direction of the hole and $\mathbf{k}$ is vertically upwards. The vector $\mathbf{j}$ is horizontal and perpendicular to $\mathbf{i}$. The origin is taken to be at the point where John hits the ball.

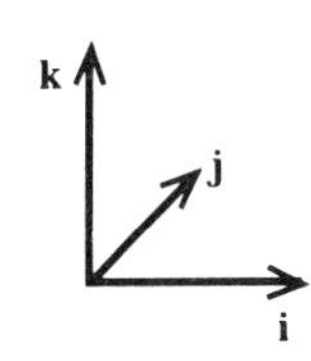

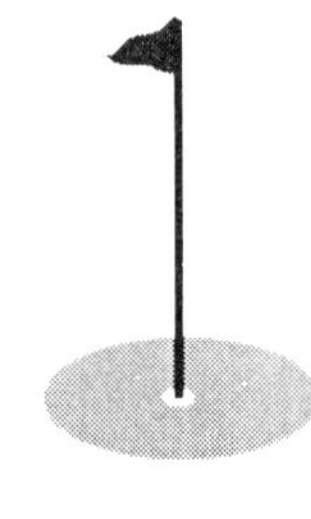

(i) Find the initial speed of the ball and the angle between its direction of flight and the horizontal at that instant.

The ball is filmed and its velocity (in metres per second) 3 seconds later is estimated to be $37\mathbf{i} - 3\mathbf{j}$.

(ii) Find the acceleration of the ball, assuming it to be constant.

(iii) Five seconds after it is hit, the position vector of the ball (in metres) is

$$187.5\mathbf{i} - 12.5\mathbf{j} + 25\mathbf{k}.$$

Show that this is consistent with the value of the acceleration you found in part (ii).

(iv) Find when and where the ball hits the ground, and its velocity and speed at that time.

(v) When John hits the ball it is travelling towards the hole but by the end of its flight it has deviated quite a long way from the correct line. How can you explain this?

15. An owl is initially perched on a tree, then goes for a short flight which ends when it dives onto a mouse on the ground. The position vector (in metres) of the owl t seconds into its flight is modelled by

$$\mathbf{r} = t^2(6 - t)\mathbf{i} + (12.5 + 4.5t^2 - t^3)\mathbf{j}$$

where the foot of the tree is taken to be the origin and the unit vectors $\mathbf{i}$ and $\mathbf{j}$ are horizontal and vertical.

(i) Draw a graph showing the bird's flight.

(ii) For how long (in seconds) is the owl in flight?

(iii) Find the speed of the owl when it catches the mouse and the angle that its flight makes with the horizontal at that instant.

(iv) Show that the owl's acceleration is never zero during the flight.

16. Ship A is 5 km due west of ship B, and is travelling on a course 035° at a constant but unknown speed v kmh^{-1}. Ship B is travelling at a constant 10 kmh^{-1} on a course 300°.

(i) Write the velocity of each ship in terms of unit vectors $\mathbf{i}$ and $\mathbf{j}$ with directions east and north.

(ii) Find the position vector of each ship at time t hours, relative to the starting position of ship A.

The ships are on a collision course.

(iii) Find the speed of ship A;

(iv) Find how much time elapses before the collision occurs.

Investigations

Umbrellas

Investigate the angle at which you should hold an umbrella for maximum protection when walking in the rain.

Investigations continued

Birds of prey

Owls and other birds of prey constantly adjust their direction of flight so that they are always heading directly towards their moving prey. Investigate the paths followed by such birds.

KEY POINTS

- The relationships between position **r**, velocity **v** and acceleration **a** may be written in vector form as

		↓ Differentiation	↑ Integration	
Position	$\mathbf{r}$			$\mathbf{r} = \int \mathbf{v}\,\mathrm{d}t$
Velocity	$\mathbf{v} = \dfrac{\mathrm{d}\mathbf{r}}{\mathrm{d}t}$			$\mathbf{v} = \int \mathbf{a}\,\mathrm{d}t$
Acceleration	$\mathbf{a} = \dfrac{\mathrm{d}\mathbf{v}}{\mathrm{d}t} = \dfrac{\mathrm{d}^2\mathbf{r}}{\mathrm{d}t^2}$			$\mathbf{a}$

- Acceleration may be due to change in direction or change in speed (or both).
- If the acceleration is constant,

$$\mathbf{v} = \mathbf{u} + \mathbf{a}t$$

$$\mathbf{r} = \mathbf{u}t + \tfrac{1}{2}\mathbf{a}t^2$$

$$\mathbf{r} = \tfrac{1}{2}(\mathbf{u} + \mathbf{v})t$$

Force

. . . the whole burden of philosophy seems to consist in this – from the phenomena of motions to investigate the forces of nature . . ..

Isaac Newton

For each of the situations below, describe the way in which the motion of the object changes and identify what causes the change in motion.

(i) A basketball being caught.
(ii) A table tennis ball being hit by a bat.
(iii) A squash ball bouncing off a wall.
(iv) A coin being dropped.
(v) A football being kicked from the penalty spot.

In all of these examples you have seen a change in motion and identified something that has caused that change. Whenever there is a change in motion there is something that has caused that change. A *force* is defined as *the physical quantity that causes a change in motion.*

Forces can

- start motion,
- stop motion,
- make objects move faster or slower,
- change the direction of motion.

In real situations several forces usually act on an object at the same time. For example in a tug of war, both teams exert considerable forces on the rope. If the rope does not move the forces exerted by each team balance each other out, so we say that the *resultant force* on the rope is zero. If one team then pulls a little harder, the resultant force will not be zero, and this will cause motion.

A similar situation arises when an aeroplane is flying in a straight line at a constant speed. The airflow over the wings provides an upward force (lift), but this is balanced by gravity which exerts a downward force. There is a forward force (thrust) provided by the engines, but this is balanced by backward forces (drag). This is shown in figure 5.1.

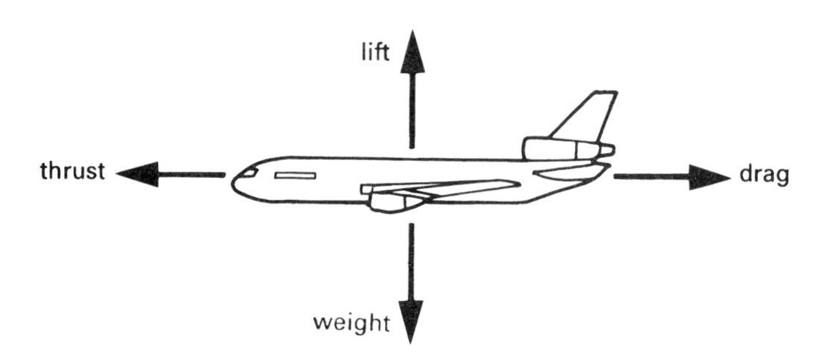

Figure 5.1

If, as in these examples, the combined forces do not cause a change in motion then the forces acting are said to be in *equilibrium*. This is the same as saying that the resultant force is zero.

For Discussion

In each of the following situations say whether the forces acting on the object are in equilibrium, by deciding whether its motion is changing.

(i) A car, that has been stationary, as it moves away from a set of traffic lights.
(ii) A motorbike as it travels at a steady 60 mph along a straight road.
(iii) A parachutist descending at a constant rate.
(iv) A box in the back of a lorry as it picks up speed along a straight, level motorway.
(v) An ice hockey puck sliding across an ice rink.
(vi) A book resting on a table.
(vii) An aeroplane flying at constant speed in a straight line, but losing height at a constant rate.

Mass

A change in motion depends not only on the force applied but also on the *mass* of the object. For example, a pram will move easily when pushed but you would find it more difficult to move a caravan. What exactly is mass?

The mass of an object is a measure of the amount of matter it contains and is the same whatever the position of the object in the Universe. The basic SI unit of mass is the *kilogram* (kg) which was originally defined as the mass of one litre of water, but that has now been replaced by a block of platinum-iridium alloy kept at Sèvres in France. The metric *tonne* is 1000 kg.

Your experience tells you that the greater the mass of an object, the harder it is to move; the smaller is its acceleration.

The force of gravity

Imagine throwing a ball straight up in the air. Once it leaves your hand it begins to slow down, until at its highest point it is at rest for an instant and then it falls down gaining speed. Throughout its flight the motion of the ball is changing and so it must be acted on by a force. The force that acts on the ball is due to gravity.

If you hold a golf ball in one hand and an athletics shot in the other, you will feel the gravitational pull even though neither object is moving. Furthermore, the effect on each hand will be different. The hand with the shot will ache long before the one with the golf ball. The force of gravity on an object depends on the mass of the object. The force of gravity acts on all objects at all times, regardless of whether they move or not.

The properties of the force of gravity can be drawn together to provide a law:

Every object on or near the Earth's surface is pulled vertically downwards by the force of gravity. The size or magnitude of the force on an object of mass M (where M is in kilograms) is Mg newtons where g is a constant whose value is approximately 9.8 ms^{-2}.

The basic unit of force is 1 newton (N). It is defined as the force needed to make an object of mass 1 kg accelerate at 1 ms^{-2}. The reason for this definition will become clearer in Chapter 7. 1000 newtons = 1 kilonewton (kN).

EXAMPLE

Calculate the force of gravity on
(i) a golf ball, mass 46 grams (ii) a shot, mass 7.26 kg

Solution

(i) Mass of golf ball = 46 grams
= 0.046 kg
Force of gravity = 0.046 × 9.8 N
= 0.45 N (to 2 sf)

(ii) Mass of shot = 7.26 kg
Force of gravity = 7.26 × 9.8 N
= 71 N (to 2 sf)

In force diagrams, the force of gravity is shown acting downwards from the centre of mass (figure 5.2).

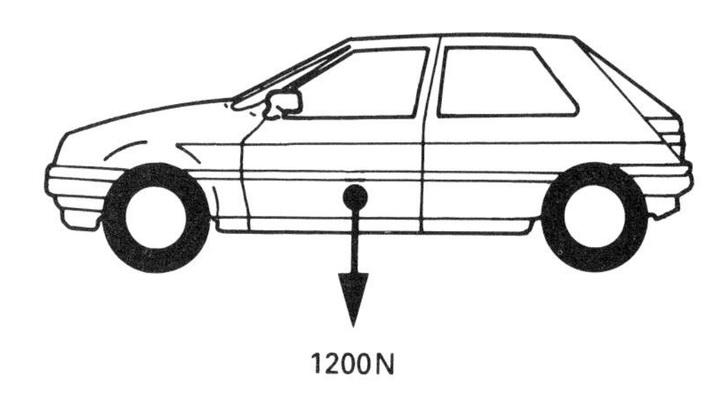

Figure 5.2

Weight

In mechanics, the word *weight* means the force of gravity acting on an object. This is different from everyday life where it is used to mean mass. For example, when a greengrocer says that the weight of a bag of apples is 1 kg this is a concise way of saying that its weight is that of a 1 kg mass. To be mathematically correct he would say that the bag of apples had a mass of 1 kg. The word weight should be used with care, to avoid confusion.

Newton's law of universal gravitation

Newton formulated a general law to describe the gravitational attraction between any two objects in the Universe. This law says that every object in the Universe attracts every other object with a force of magnitude

$$F = \frac{Gm_1 m_2}{d^2}$$

where m_1 and m_2 are the masses of the objects in kilograms, d is the distance between the centres of the objects in metres and G is the *gravitational constant*, which is 6.67×10^{-11} in S.I. units.

The mass of the Earth is 5.98×10^{24} kg and its radius is 6.37×10^6 m, so the force on an object of mass M on the earth's surface is

$$F = \frac{(6.67 \times 10^{-11}) \times (5.98 \times 10^{24})M}{(6.37 \times 10^6)^2}$$

$$= 9.8\,M \text{ (in newtons).}$$

The law of universal gravitation is therefore consistent with the definition of the force of gravity given earlier, but it provides a much more general model. It can be applied to any two objects.

Notice that the gravitational force between two bodies acts equally on both bodies as shown in figure 5.3. So not only is there a force on you towards the Earth, but also an equal force on the Earth towards you.

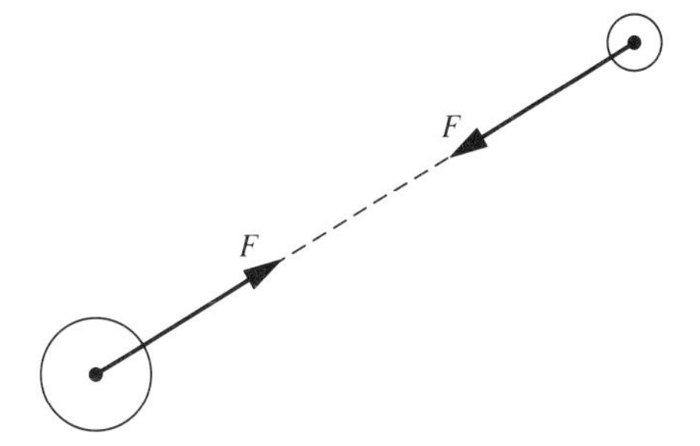

Figure 5.3

Exercise 5A

Data for Exercise 5A: g on Earth $= 9.8\,\text{ms}^{-2}$; g on Moon $= 1.6\,\text{ms}^{-2}$

1. A person has mass 65 kg. Calculate the force of gravity
   (i) of the Earth on the person, (ii) of the person on the Earth.

2. Two balls of the same shape and size but with masses 1 kg and 3 kg are dropped from the same height. Which hits the ground first? If they were dropped on the Moon what difference would there be?

3. A ball has been thrown straight up in the air. The diagram shows the ball at three positions, A on the way up, B at the top and C on the way down.
   (i) Copy the diagram, indicating the force of gravity on the ball at each position.
   (ii) Do any other forces act on the ball?

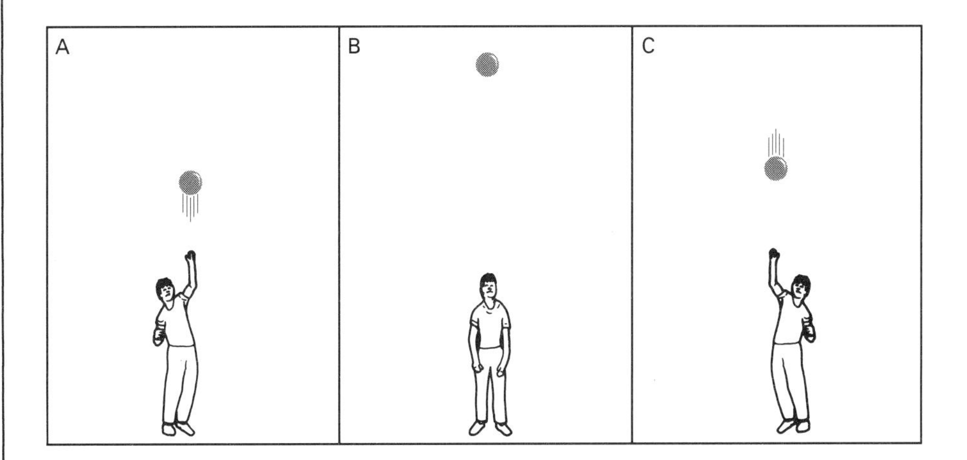

4. (i) What is your mass?
   (ii) Calculate your weight when you are on the Earth's surface.
   (iii) What would your weight be if you were on the Moon?
   (iv) When people say that a baby weighs 4 kg, what do they mean?

Exercise 5A continued

5. Calculate the magnitude of the force of gravity on the following objects and draw a diagram showing its direction:
(i) a car of mass 1.2 tonnes
(ii) a letter of mass 50 grams
(iii) a suitcase of mass 15 kg

6. Estimate the force of gravity on the following objects:
(i) a table tennis ball
(ii) a car
(iii) an apple.

Tension forces

When a light string is pulled, the string will exert a *tension* force opposite to the pull. The tension force acts along the string, and is the same throughout its length.

Figure 5.4 shows a string supporting an object of mass *m* which is at rest. The force of gravity pulls down on the object, but the string exerts an upward *tension* force of equal magnitude to balance the downward force.

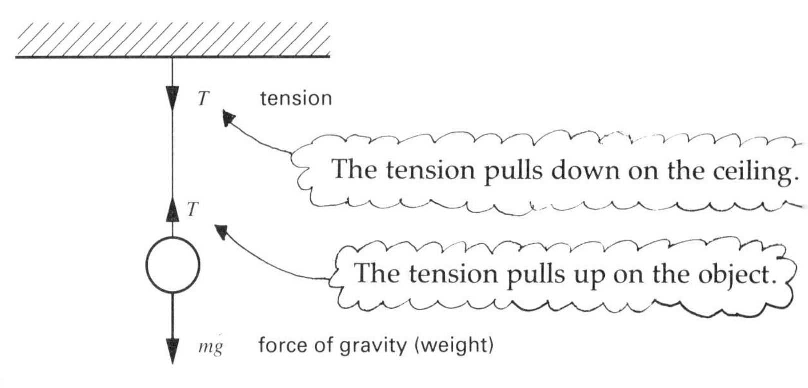

Figure 5.4

The tension in the string exerts an upward force on the object which is represented by an upward arrow. The force of gravity exerts a downward force on the object which is represented by a downward arrow. Both forces should be shown on a force diagram of the object. Note that the string exerts an equal tension force on the support which would be shown on a force diagram of the support.

EXAMPLE

Calculate the tension in the strings of the following systems, both of which are at rest.

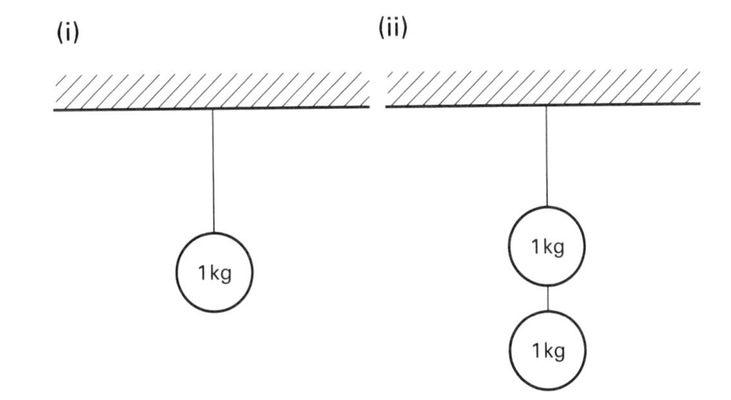

Solution

(i) Force of gravity on mass $= 1 \times 9.8\,\text{N} = 9.8\,\text{N}$.

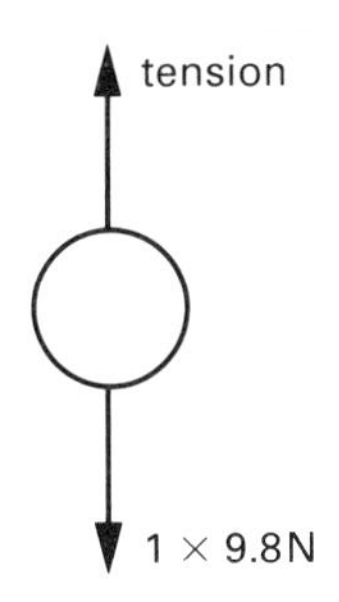

Since the mass is in equilibrium, the tension must act upwards with the same size force, so

$$\text{tension} = 9.8\,\text{N}.$$

(ii) Force of gravity on each mass $= 1 \times 9.8 = 9.8\,\text{N}$.
Considering first the lower mass, on which two forces act, as the mass is at rest the tension (T_1) is equal to the force of gravity, so,

$$T_1 = 9.8\,\text{N}.$$

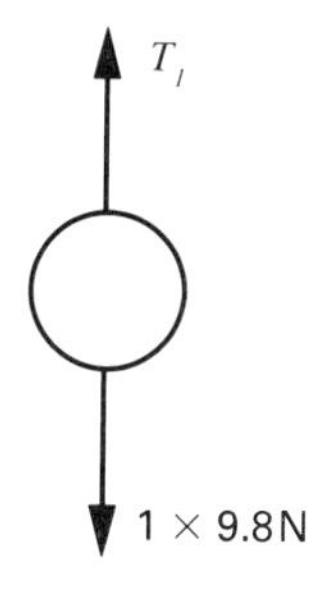

Three forces act on the top mass: the two tensions and the force of gravity. The tension in the top string T_2 must balance both the other forces.

$$\begin{aligned} T_2 &= T_1 + 9.8 \\ &= 9.8 + 9.8 \\ &= 19.6\,\text{N}. \end{aligned}$$

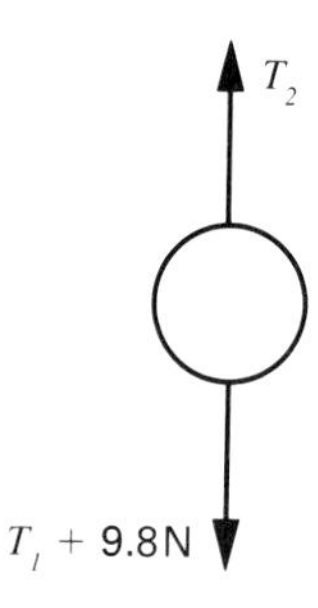

EXAMPLE

The diagram shows a man holding the end of a rope that passes over a smooth pulley and supports a crate of mass 50 kg. If the crate is at rest find the force exerted by the man on the rope.

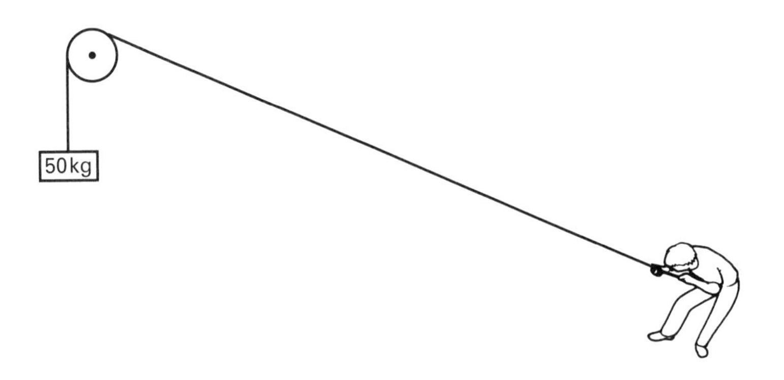

Solution

Two forces act on the crate, the tension in the rope and the force of gravity. These are shown on the force diagram below.

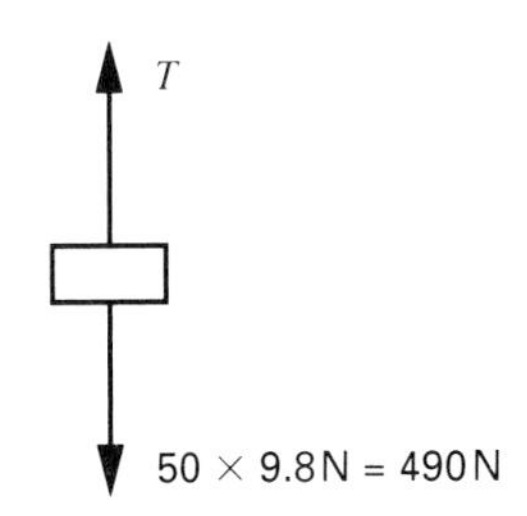

As the box is at rest the tension must balance the force of gravity, so the tension in the rope is 490 N. The tension is the same throughout its length, so the force exerted by the man on the rope is 490 N.

Notice that in the last example the pulley was smooth, so it exerted no force along the rope. If the pulley were not smooth, the tension might be different either side of it.

Exercise 5B

In this exercise you may assume that all strings are light and all pulleys are smooth.

1. The diagram shows a mass suspended on a string.
   (i) Show on a diagram the forces acting on the mass.
   (ii) Find the tension in the string if the mass is at rest.

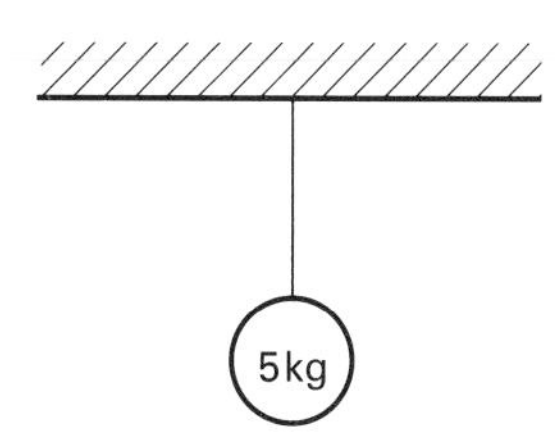

2. The diagram shows two objects of mass 0.1 kg suspended from a string which passes over a pulley. The system is at rest.

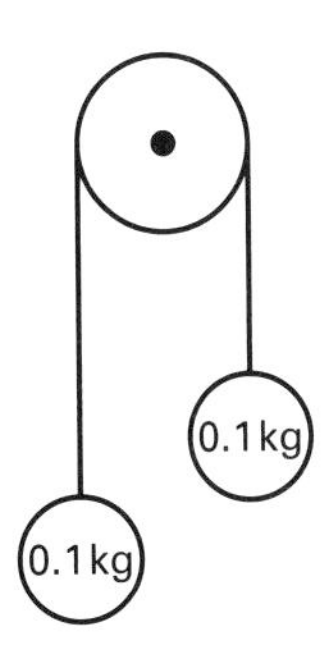

   (i) Show on a diagram the forces acting on each mass.
   (ii) What is the magnitude of the tension in the string if the system is at rest?

3. A gymnast of mass 50 kg hangs at rest from a bar. Find the tensions (assumed equal) in each of her arms.

Exercise 5B continued

4. The diagram shows a boat of mass 650 kg suspended at rest using a rope passing over a pulley.

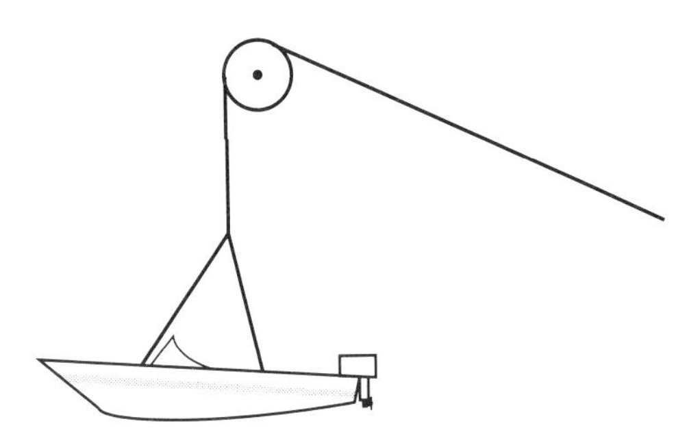

(i) Draw a diagram to show the forces acting on the boat.
(ii) What is the tension in the rope passing over the pulley?

5. What tension must the string shown below be able to withstand if the 10 kg mass is to remain at rest?

6. The system shown below is at rest.

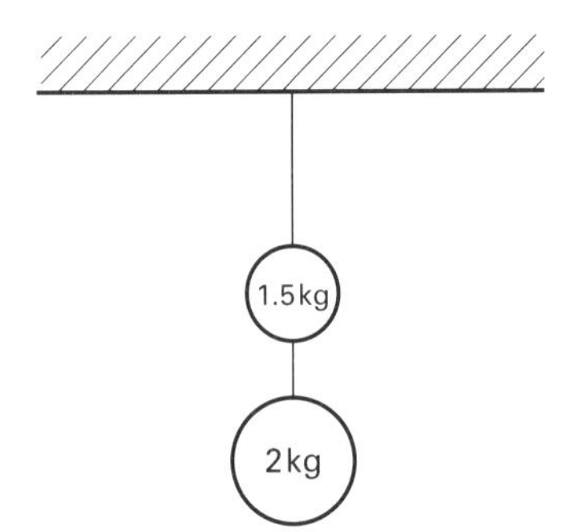

(i) Draw diagrams to show the forces acting on each mass.
(ii) Find the tension in the lower string.
(iii) Find the tension in the upper string.

7. For the system shown below, find the tension in each string. Assume that the top strings share the load equally.

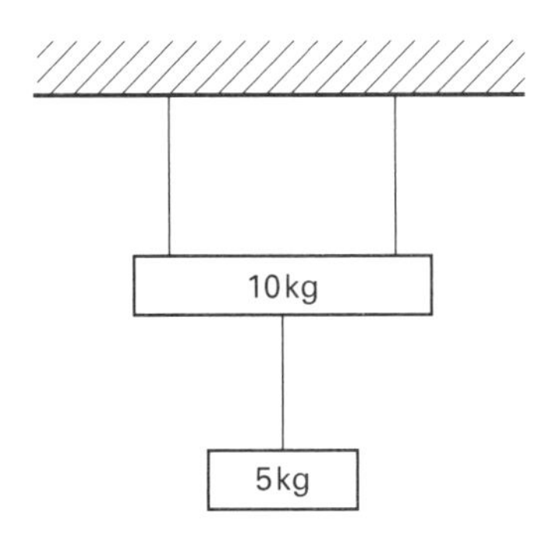

8. A conker is swung in a vertical circle as shown in the diagram.

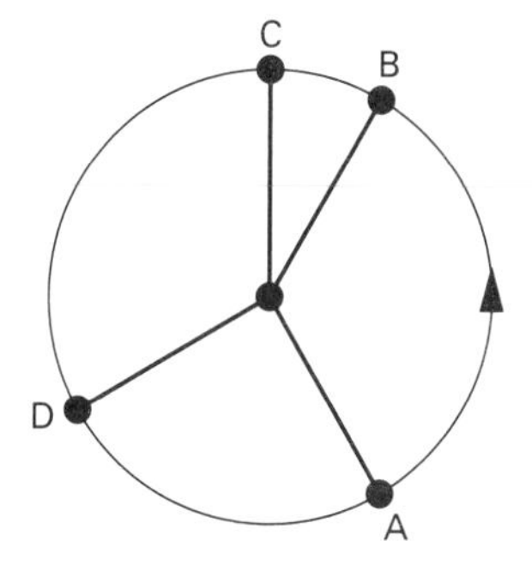

(i) Draw diagrams to show the force of gravity and the tension acting at each point.

(ii) Are there any other forces acting?

Thrust

A rod, which is rigid, can exert tension forces in a similar way to a string when it is used to support or pull an object, as shown in figure 5.5.

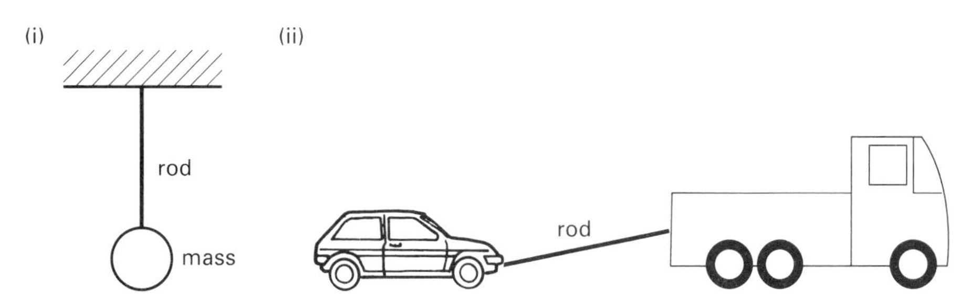

Figure 5.5

For Discussion

When the breakdown truck in figure 5.5 begins to pull the car, the rod exerts a tension force on the car. A rope would behave in the same way in this situation.

Discuss what happens to (i) a rope, (ii) a rod, if the truck slows down.

As well as tension forces, rods can also exert thrust forces when they are in compression. Each leg of a table exerts an upward thrust force on the top of the table, and a downward thrust force on the floor (figure 5.6). The thrust forces act upwards on the table top to maintain equilibrium, balancing the force of gravity, which acts downwards on the table top.

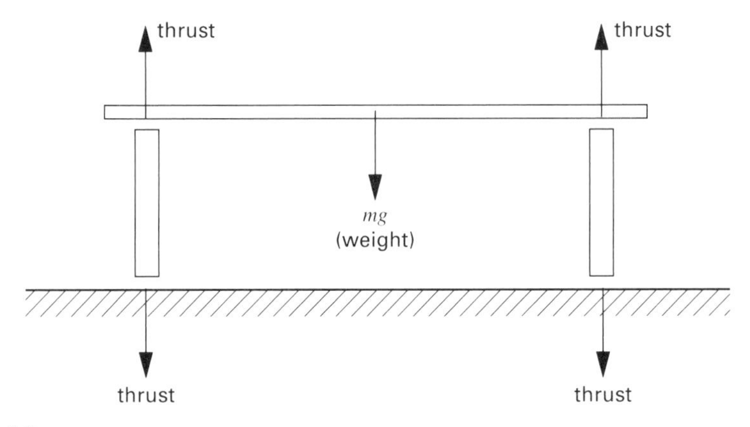

Figure 5.6

For Discussion

In each of the following situations, discuss whether the rod is exerting a tension or a thrust.

(i) The rod that forms part of a clock pendulum.
(ii) A vertical pole in a ridge tent.
(iii) The telegraph pole shown in diagram (a).
(iv) The rods forming a step-ladder as in diagram (b).

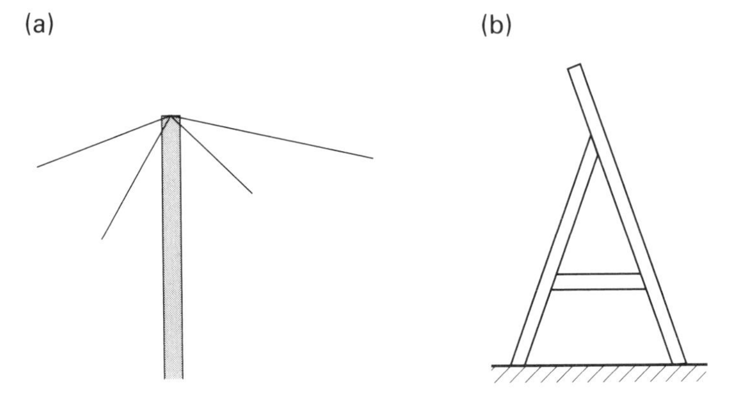

Contact forces

Normal reaction

Think about a book on a table. The force of gravity acts downwards on it. As the book is in equilibrium, at rest, there must also be an upward force acting on it. The table exerts an upward force on the book. This upward force is called a *normal reaction*. It is called normal because its line of action is normal (at right angles) to the surface of the table. Since the book is in equilibrium, the normal reaction must be equal and opposite to its weight (figure 5.7).

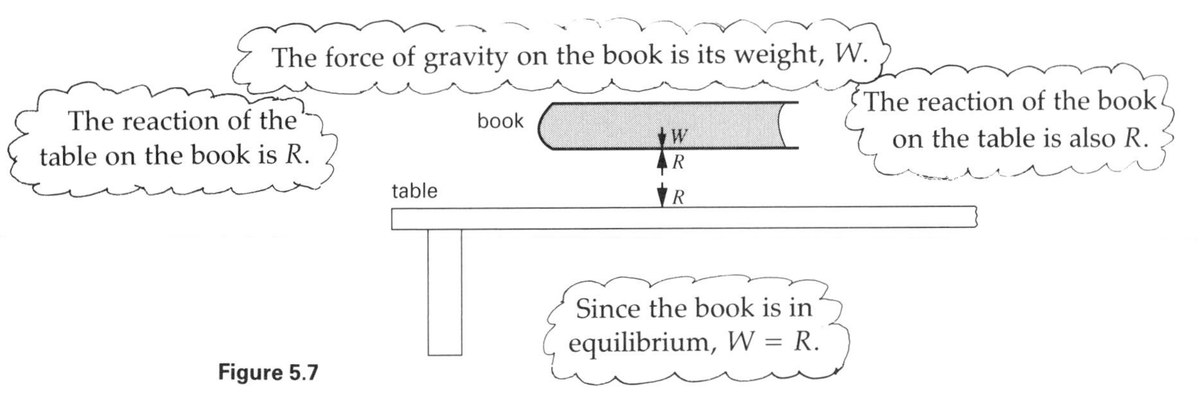

Figure 5.7

Note that the book exerts a downward force on the table, also equal to its weight.

Friction

Place a heavy book on a table and push it very lightly with your finger. Nothing happens. The force from your finger is balanced by an equal frictional force in the opposite direction (figure 5.8).

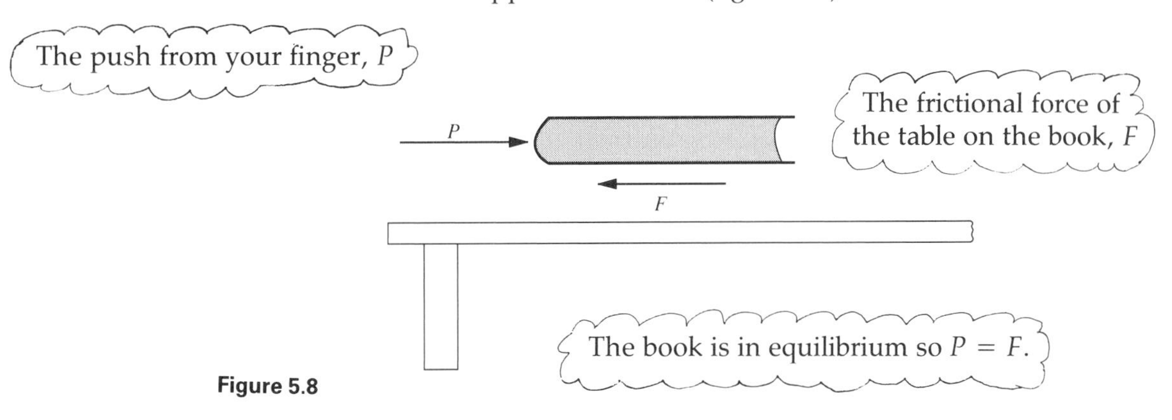

Figure 5.8

Now increase the force P with which your finger is pushing the book. As P increases, so does the frictional force F opposing it. They balance each other, so $P = F$.

Until . . . the book moves. At that point the frictional force F has reached the greatest value it can take, and it is no longer able to balance P.

A frictional force will always act in the direction opposed to motion. If an object is moving, the frictional force will take its greatest possible value. A frictional force never causes motion; it may prevent motion, or it may slow down something that is moving.

The frictional force acts parallel to the surface of the table, and so it is at right angles to the normal reaction as shown in figure 5.9.

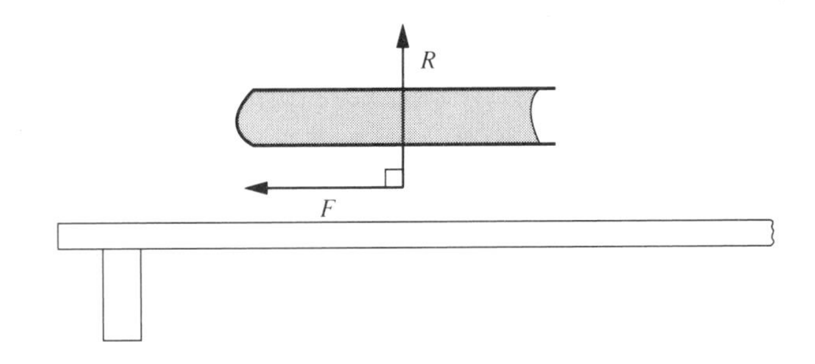

Figure 5.9

The total contact force of the table is made up of two parts: the normal reaction at right angles to the surface, and the frictional force parallel to the surface.

When modelling, the word *smooth* is used to mean that there is no friction. You met a 'smooth' pulley on p. 52.

A model for friction will be explored in *Mechanics 2*.

EXAMPLE

A ladder leans against a wall. On a diagram show
(i) the force of gravity,
(ii) the friction forces,
(iii) the normal reactions.

Solution

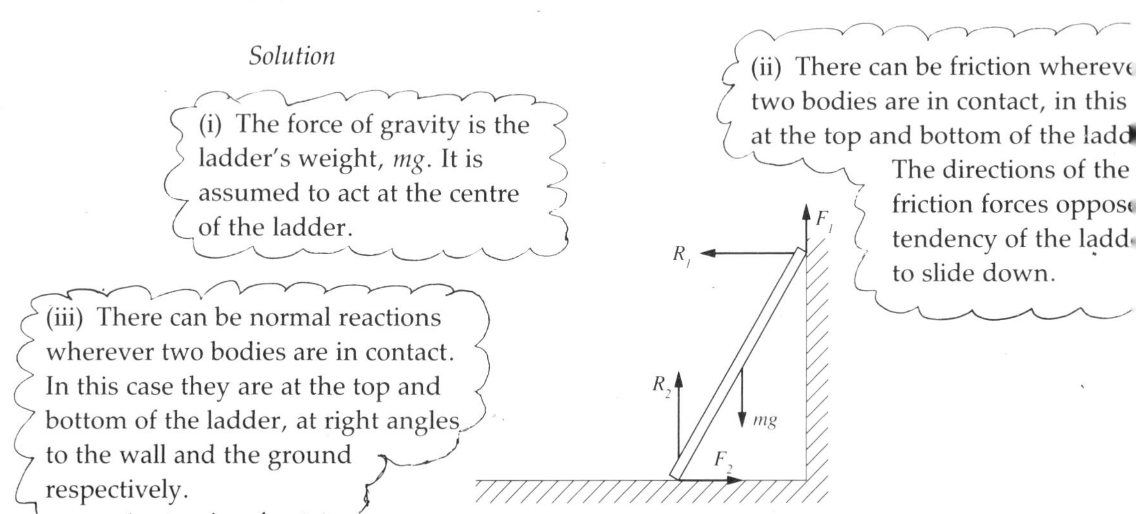

EXAMPLE

A child pulls a sledge across a snow-covered field at a constant speed. If the child exerts a horizontal force of 12 N on the sledge, calculate the magnitude of the friction force present.

Solution

The diagram shows the forces on the sledge. As the sledge is moving at a constant speed the forces acting on it must be in equilibrium. This means that the resultant horizontal force is zero, so $F = 12$ N.

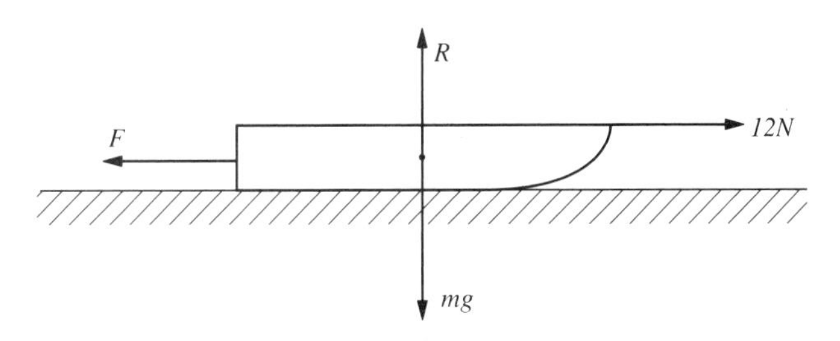

Other types of force

There are several other ways in which a force can act on an object.

Air resistance

This force acts in a direction opposite to the motion of the object. The same is true for resistance in a liquid.

Upthrust/buoyancy

When an object is immersed in a fluid it receives an upthrust equal to the weight of the fluid displaced.

Electrical and magnetic forces

Objects may be subject to forces due to electricity or magnetism, but this is outside the scope of this book.

Exercise 5C

If you can draw correct force diagrams, you are well on your way to solving most mechanics problems. In this exercise you are asked to draw force diagrams using the various types of force you have met in this chapter.

Remember that forces are not caused by motion.

Exercise 5C continued

1. For each of the following situations, draw the force diagram(s) for the object(s) named in italics.

(i)

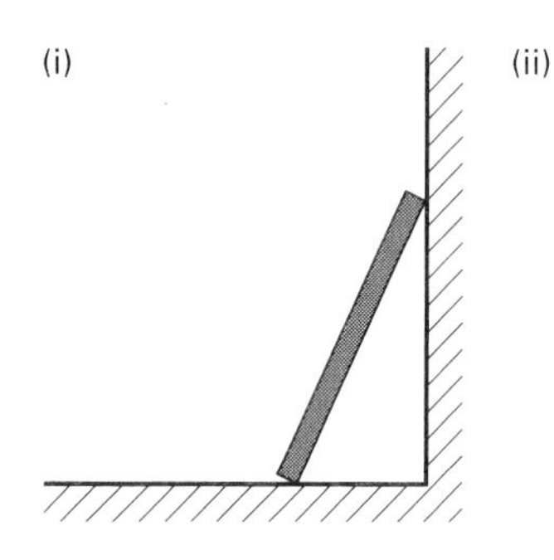

A *paving stone* leaning against a wall.

(ii)

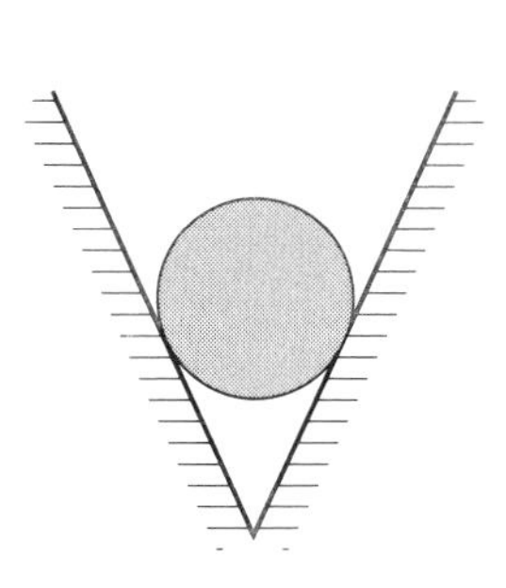

A *cylinder* at rest on smooth surfaces.

(iii)

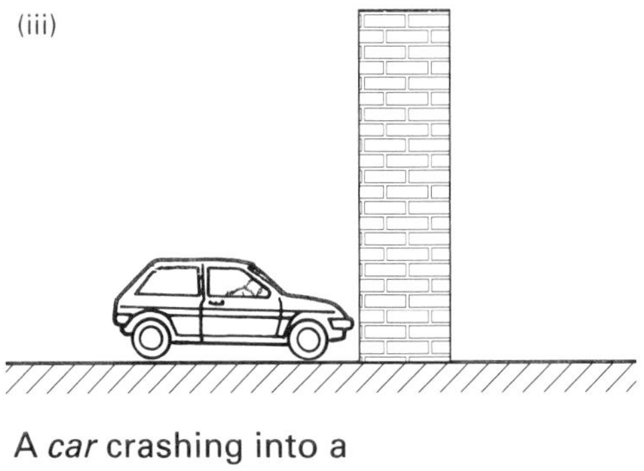

A *car* crashing into a wall.

(iv)

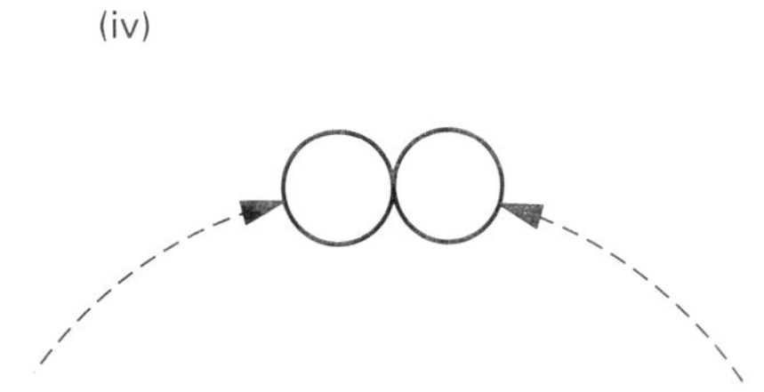

Two balls colliding in mid-air

(v)

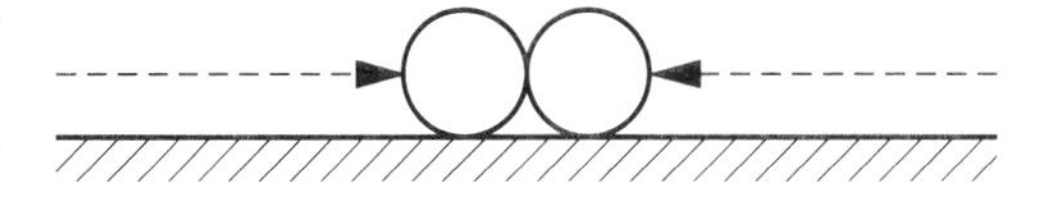

Two balls colliding on a snooker table.

2. A cricket ball follows the path shown below. Draw arrows to indicate the forces acting on the ball at the positions shown (include air resistance).

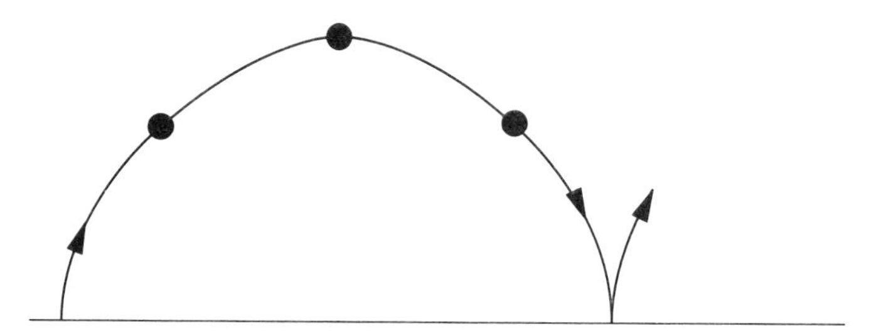

9·8

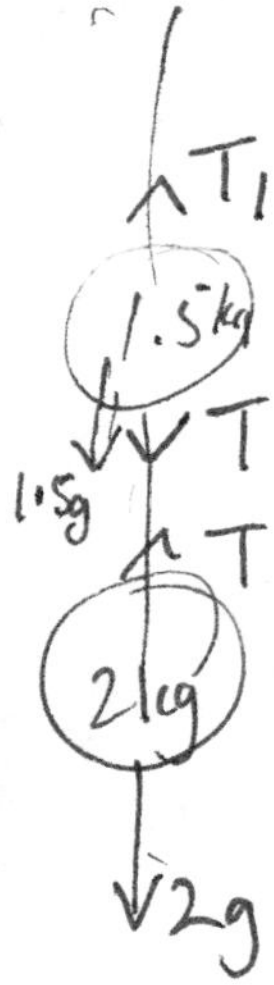

T_1

1.5kg

1.5g

T

T

2kg

2g

$T = 2g. = 18.6$ N

$T_1 = 3.5g = 34.3$ N

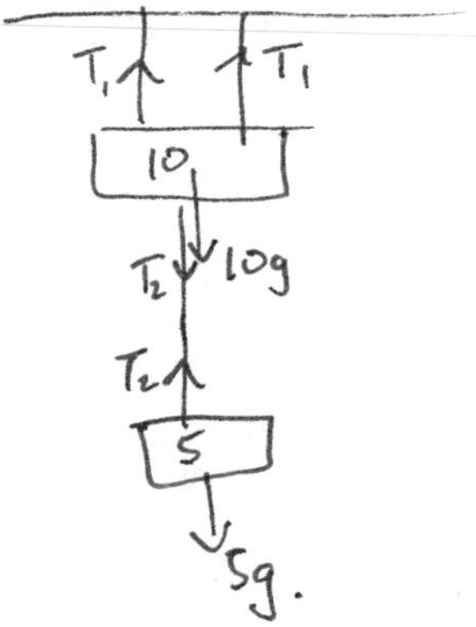

T_1 T_1

10

T_2 10g

T_2

5

5g.

$T_2 = 5g = \underline{49\text{ N}}$

$T_2 + 10g = 2T_1$

$49 + 98 = 2T_1$

$T_1 = \frac{147}{2} = \underline{73\frac{1}{2}\text{ N}}$

1xel Paveley

Peltier Spink

$f(x) = x^5 - 5x + 3$

$f(1) = 1 - 5 + 3 =$

Plattf Benth.

Wilson

$x^5 - 5x + 3 = 0.$

3

-2 -1 1 2

[-2, -1]. [0, 1] [1, 2]

Geog Thurs 16 Oct.

$\frac{9}{4}$

$\frac{400}{50} = 8$

Stringer Dentist

$.95 \times 8$

$\frac{400}{.95 \times 8}$

x 0 0.1 0.2

y

$\frac{7x}{1-2x}$

$\frac{(1-2x)7 - 7x(-2)}{(1-2x)^2}$

$\frac{7 - 14x + 14x}{(\quad)^2}$

$\frac{8x-1}{3x+2}$

$\frac{5x^3}{1-3x^2}$

$\frac{(1-3x^2)15x^2 - 5x^3(-6x)}{(1-3x^2)^2}$

$\frac{5x^2\{3 - 9x^2 + 6x^2\}}{(\quad)^2}$

$\frac{5x^2 \cdot 3(1-x^2)}{(\quad)}$

[0.618, 0.619]

[0.6180, 0.6181]

Root is 0.61805

$\pm .00005$

max^m error of

0.618 to 3dp.

Solution bounds .6180, .6181.

$\uparrow 20 \quad \downarrow g$ ie $\uparrow -g$

h

$v^2 = u^2 + 2as$

$0 = 20^2 - 2gh$

$h = \frac{20 \times 20}{20} = \underline{20}$

$v = u + at$

$0 = 20 - gt$

$t = 20/g$

$s = 0$ ~~$s = ut + \frac{1}{2}$~~

$v^2 = u^2 + 2as$

$v^2 = 20^2$

$v = 20 \downarrow$

$\frac{3}{5} \cdot 20 = 12$

$v^2 = u^2 + 2as$

$0^2 = 12^2 + 2gh$

$h = \frac{12 \times 12}{2g}$

$v = u + at$

$0 = 8 - gt$

$\uparrow 34.3$ 49

$s = ut + \frac{1}{2}at^2$

$49 = 34.3t - 4.9t^2$

$490 = 343t - 49t^2$

$70 = 49t - 7t^2$

$10 = 7t - t^2$

~~$t = 3$~~

$t^2 - 7t + 10 = 0$

$(t - 5)(t - 2) = 0$

2 5

3

.7

$\uparrow$

9.6

$s = ut + \frac{1}{2}at^2$

$-19.6 = 14.7t - \frac{9.8}{2}t^2$

$-196 = 147t - 49t^2$

$7t^2 - 21t - 28 = 0$

$\underline{t^2 - 3t - 4 = 0}$

$+0$ 1 sec later $\downarrow 11$

15.04 kg
10 kg
7.64 kg

h.

$s = ut + \frac{1}{2}at^2$

$h = \frac{1}{2}gt^2$

$h = 5t^2$

$s = ut + \frac{1}{2}at^2$

$h = 11(t-1) + \frac{1}{2}g(t-1)$

$h = 11t - 11 + 5(t^2 - 2t + 1)$

$h = 5t^2 + t - 6$

~~$\frac{1}{2}gt$~~ $5t^2 = 5t^2 + t - 6$

$t = 6$

$h = 5 \times 36$

$= 180$

18 18 20 mg

$16\cos 20 = 15.0350819326$

15 N

T_2 T_1 mg

15 N

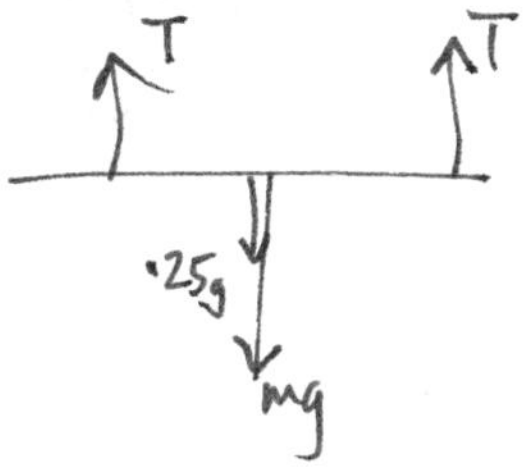

T T .25g mg

$2T = 15.285$

$T = 7.643$ N

$15 = mg$

$m = 15/g = 1.5$ kg.

$T \simeq 3.8$

3. Draw labelled diagrams showing all the forces acting on the following objects.
   (i) A car towing a caravan.
   (ii) A caravan being towed by a car.
   (iii) A person pushing a supermarket trolley.
   (iv) A suitcase on a horizontal moving floor (as at an airport).
   (v) A chair with a person sitting on it.
   (vi) A sledge being pulled uphill.

4. Ten boxes each of mass 5 kg are stacked on top of each other.
   (i) What forces act on the top box?
   (ii) What forces act on the bottom box?

5. On the Moon the acceleration due to gravity is $1.6\ \text{ms}^{-2}$. What reaction force would an astronaut of mass 70 kg experience while standing on the Moon?

6. The diagrams show a box of mass m under different systems of forces. In both situations, the box is at rest. State the value of F_1 and F_2, the friction forces acting.

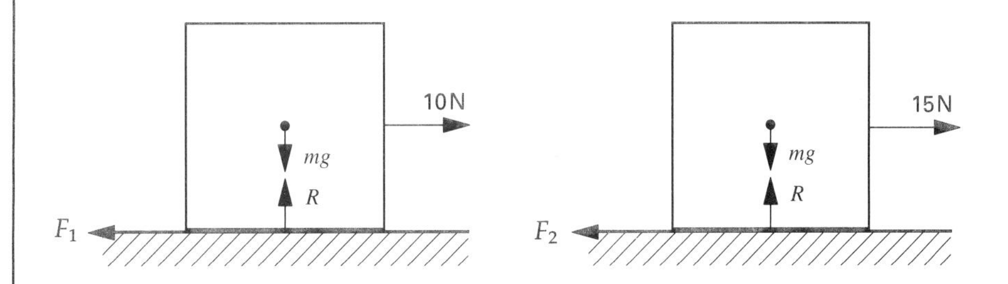

   In one of the situations the box is on the point of slipping. State the maximum value of the friction.

7. In this diagram the pulley is smooth and the 2 kg mass is on the point of slipping.

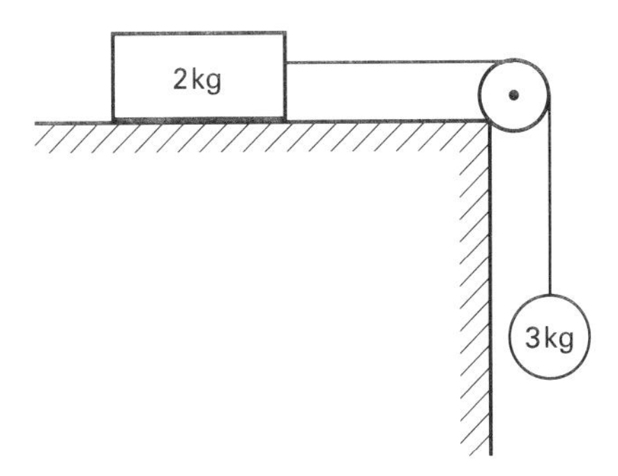

   (i) Draw diagrams to show the forces acting on each mass.
   (ii) Find the tension in the string.
   (iii) Calculate the magnitude of the friction force on the 2 kg mass.

Exercise 5C continued

8. In this diagram the pulleys are smooth and the 4 kg mass is on the point of slipping.

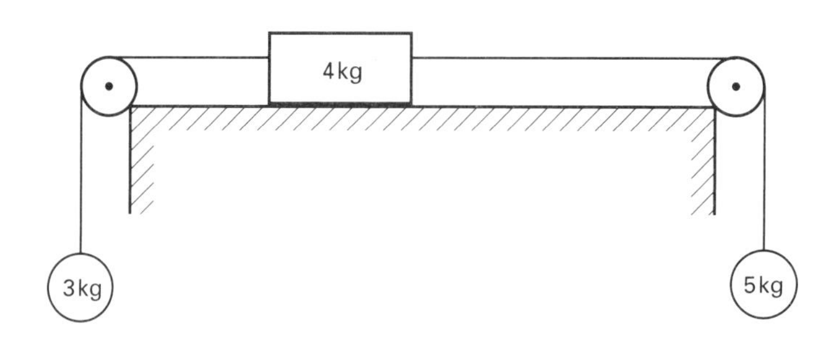

(i) Show the direction of the friction force on the 4 kg mass.
(ii) Calculate the magnitude of the friction.

9. The diagram shows a simple model of a crane. The structure is at rest in a vertical plane. The rod and cables are of negligible mass and the load suspended from the joint at A is 30 N.

Draw a diagram showing the forces acting on
(i) the load, (ii) the joint at A.
Is the rod in compression or tension?

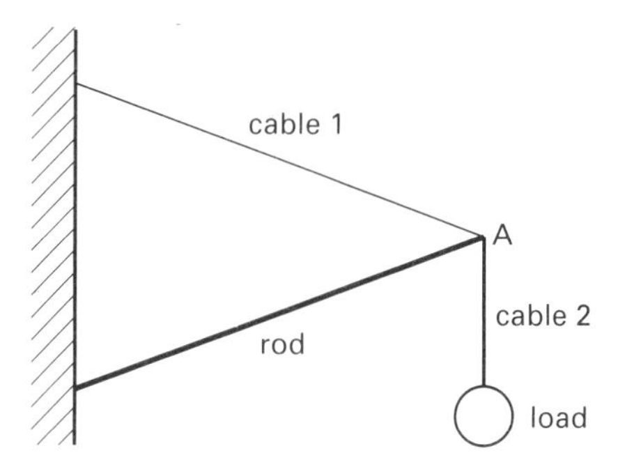

10. Which of these external forces act on the bodies shown in the following photographs?
A force of gravity
B friction between solids
C resistance forces in liquids
D air resistance
E upthrust (due to displacement of a gas or liquid)
F normal reaction
Indicate roughly where the forces act.

Investigations

Buoyancy

Devise an experiment to investigate how the buoyancy or upthrust on a ball varies as it is submerged in water.

Sporting reactions

Investigate how athletes make use of reaction forces. For example, consider weight lifters, or high jumpers.

KEY POINTS

- A force is a physical quantity that causes a change in motion.
- Force is a vector quantity.
- There are many types of force including:
  - weight (the force of gravity on an object)
  - tension
  - thrust
  - normal reaction
  - friction
  - air resistance
  - upthrust/buoyancy
  - electrical and magnetic forces.

Before starting work on a problem, draw a diagram and mark in the forces involved.

6 Modelling forces with vectors

One must learn by doing the thing; though you think you know it, you have no certainty until you try.

Sophocles

The photograph shows a crane on a building site. What is the greatest mass it can lift? How strong do the cables need to be?

In situations like this many forces act in a variety of directions and we need an approach to deal with them. Chapter 5 introduced the main types of force that you are likely to encounter. This chapter shows how vectors can be used for describing and manipulating forces. You have seen already how a force acting in one direction can be balanced by a force in the opposite direction, but a more formal method is needed when the forces act in a variety of directions.

Several forces acting at a point

When several forces act at the same point, you can replace them with a single equivalent force, called their *resultant*. Figure 6.1 shows four forces acting at the same point, P. To find their resultant you add them, using the vector addition methods which you met in Chapter 3.

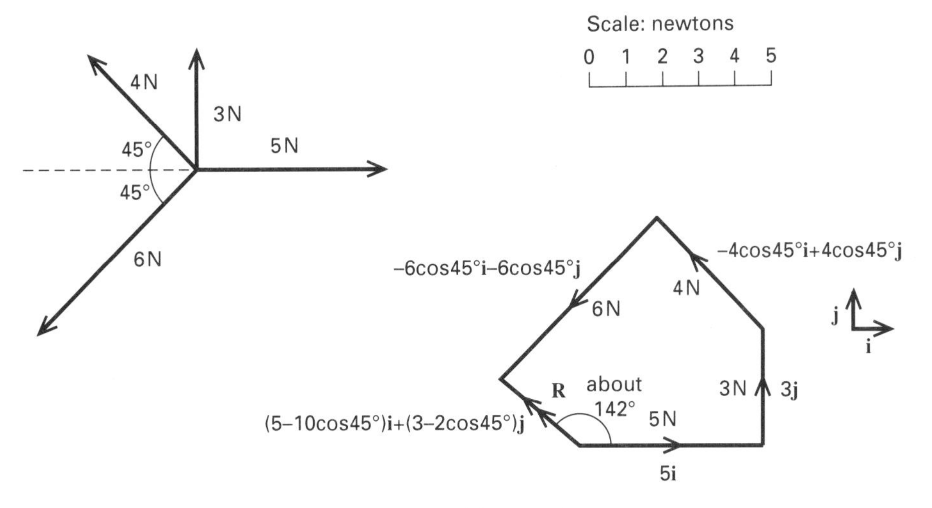

Figure 6.1

Figure 6.2

From the scale drawing in figure 6.2 you can see that the resultant makes an angle of about 142° from the direction of the 5 N force, and has magnitude about 2.6 N.

An alternative method of finding the resultant is to calculate it using components. The unit vectors **i** and **j** are defined to be in the directions shown in the diagram. This approach allows the resultant to be found precisely. The component forms of the four forces are shown on the diagram.

Adding the four vectors gives

$$\begin{aligned}\mathbf{R} &= (5\mathbf{i}) + (3\mathbf{j}) + (-4\cos 45°\mathbf{i} + 4\cos 45°\mathbf{j}) + (-6\cos 45°\mathbf{i} + 6\cos 45°\mathbf{j}) \\ &= -2.07\mathbf{i} + 1.59\mathbf{j}.\end{aligned}$$

From figure 6.3, its magnitude is given by

$$|\mathbf{R}| = \sqrt{(-2.07)^2 + 1.59^2} = 2.61.$$

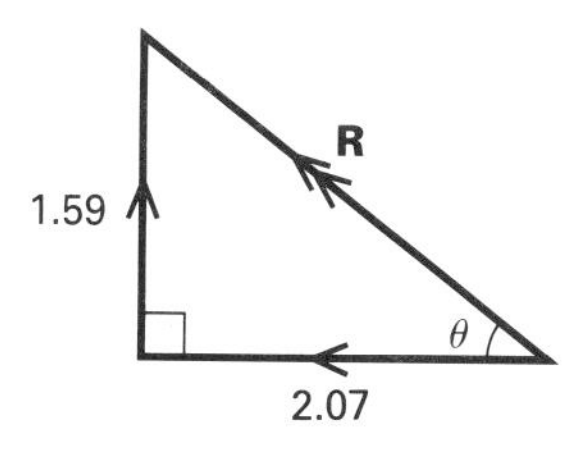

Figure 6.3

The angle θ is given by $\tan\theta = \dfrac{1.59}{2.07}$

$$\text{so } \theta = 37.5°$$

and the angle **R** makes with the **i**-direction is $180° - 37.5° = 142.5°$.

NOTE *You could have obtained the same result working with column vectors as follows.*

$$\begin{pmatrix}5\\0\end{pmatrix}+\begin{pmatrix}0\\3\end{pmatrix}+\begin{pmatrix}-4\cos 45^\circ\\-6\cos 45^\circ\end{pmatrix}+\begin{pmatrix}-6\cos 45^\circ\\-6\cos 45^\circ\end{pmatrix}=\begin{pmatrix}-2.07\\+1.59\end{pmatrix}.$$

You can use either notation, as they both lead to the same results.

Equilibrium

If the forces acting at a point exactly balance, that is if their resultant is zero, they are said to be in *equilibrium*.

When you draw a vector diagram to show such forces added together, the final point is the same as the starting point. It is a closed figure, as shown by the two examples in figure 6.4.

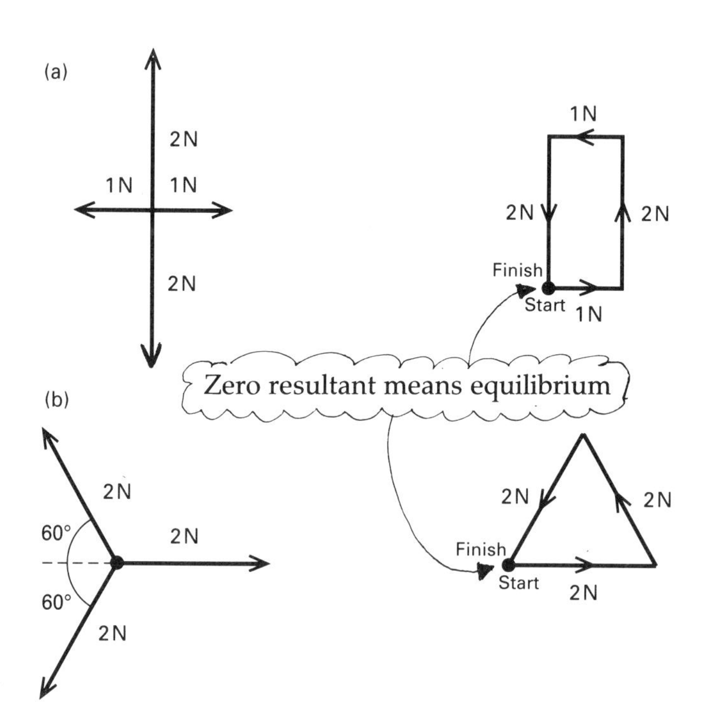

Figure 6.4

This result is sometimes stated as the polygon of forces theorem:

If several forces acting on a point are in equilibrium, then they can be represented in size and direction by a closed polygon.

Experiment The diagram shows two strings tied to a ring at A and passing over smooth pulleys at B and C. Masses M_1 and M_2 are suspended from the free ends of the strings, and mass M_3 is suspended from the ring at A. The whole system is mounted on a vertical board. Set up the apparatus as shown.

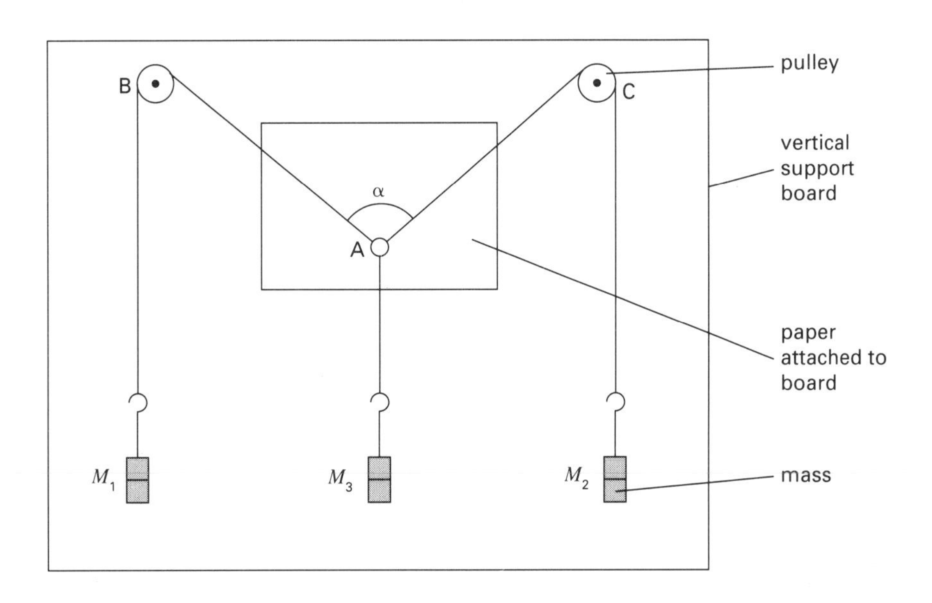

1. Before doing any experiments, try to answer these questions intuitively.
   - Do the lengths of the strings affect the angles?
   - For any given set of three masses, does the system have more than one equilibrium position?
   - Is there a combination of masses for which $\alpha = 90°$? Or for which $\alpha = 180°$?
   - Is it always possible to find an equilibrium position?

   Now use the apparatus to investigate these questions.

2. The following procedure allows you to demonstrate the way in which forces combine.
   - State which three forces act on the ring at A.
   - The forces on the ring are in equilibrium. Use this to calculate the tension $\mathbf{T}_1$ in the left hand string. Use a similar method to find the tension $\mathbf{T}_2$ in the right hand string.
   - Mark points A, B and C on the paper behind, and draw AB and AC. Using a suitable scale, mark distances along AB and AC to represent the tensions $\mathbf{T}_1$ and $\mathbf{T}_2$ in the strings.
   - The two tensions, $\mathbf{T}_1$ and $\mathbf{T}_2$, are equivalent to a single resultant force, which must act straight upwards to balance the force of gravity on mass M_3. If your resultant does not act vertically upwards, check your working.

Exercise 6A

For each of the situations below, carry out the following steps.

(i) Draw a scale diagram to show the vector sum of the forces.

(ii) State whether you think the forces are in equilibrium and, if not, *estimate* the magnitude and direction of the resultant.

(iii) Write the forces in component form, using the directions indicated, and perform the vector addition in this form and so obtain the magnitude and direction of the resultant.

(iv) Compare your answers to parts (ii) and (iii).

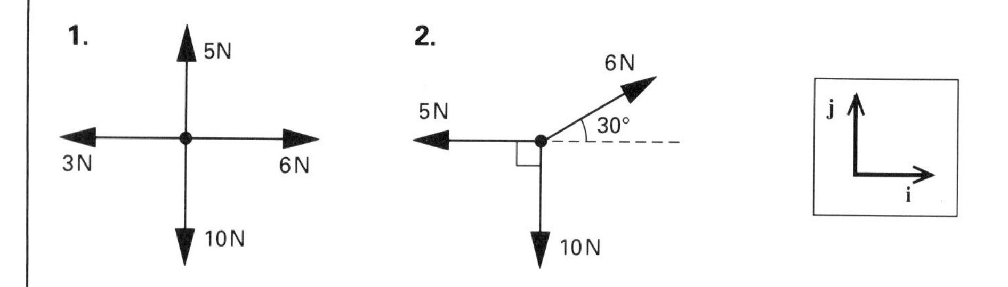

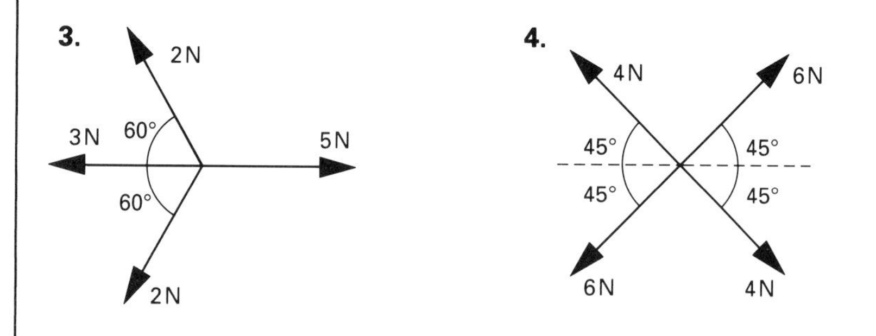

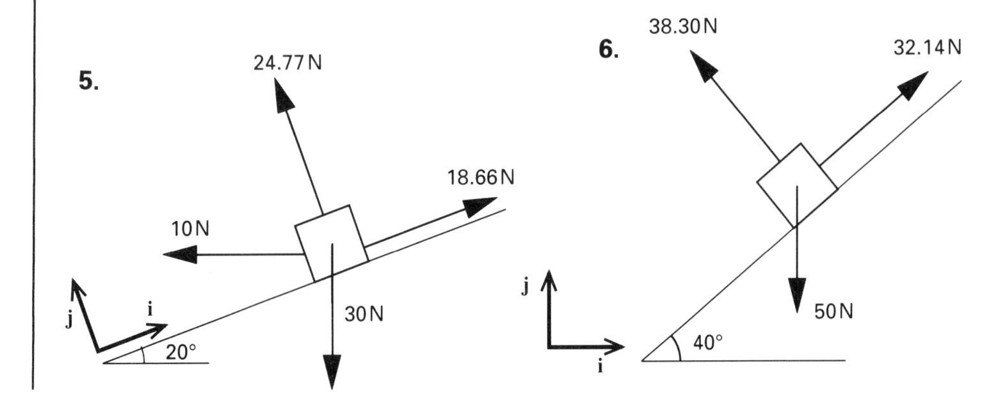

Modelling real situations

There are many situations in everyday life involving several forces acting at a point, as shown in the following examples and in Exercise 6B.

EXAMPLE

A sign of mass 10 kg is to be suspended by two strings arranged as shown in the diagram below. Find the tension in each string.

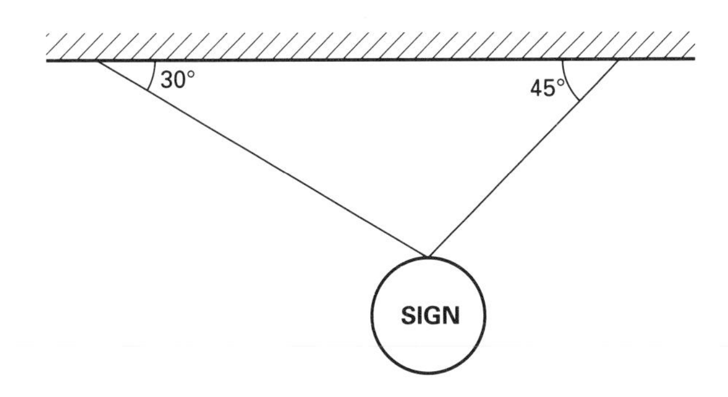

Solution

The force diagram for this situation is given below

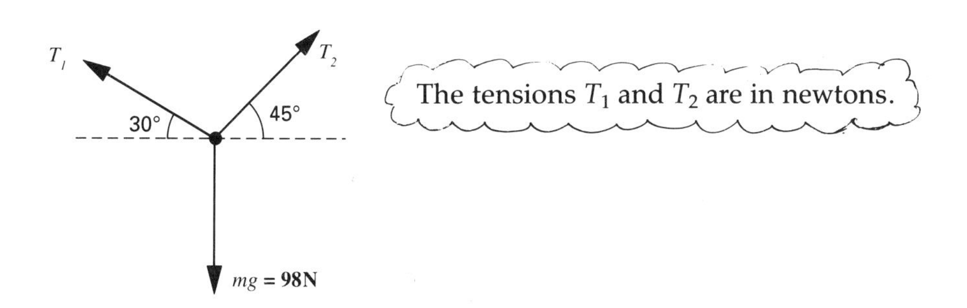

There are several ways you can proceed.

Method 1 (Resolving forces)

Vertically (↑): $T_1 \sin 30° + T_2 \sin 45° = 98 \Rightarrow \quad 0.5T_1 \quad + 0.707T_2 = 98$

Horizontally (→): $-T_1 \cos 30° + T_2 \cos 45° = 0 \Rightarrow -0.866T_1 + 0.707T_2 = \quad 0$

Subtracting $\quad 1.366T_1 = 98$

This gives

$$T_1 = \frac{98}{1.366} = 71.7.$$

Back-substitution gives $T_2 = 87.9$

The tensions are 71.7 N and 87.9 N.

Method 2 (Scale drawing)
Since the 3 forces are in equilibrium, they can be represented by a closed 3-sided polygon, in other words a triangle.

The first step is to draw a vertical line to represent the weight, 98 N. Then add the force T_2 at 45° to the horizontal (note the length of this vector is unknown), and the force T_1 at 30° to the horizontal (60° to the vertical).

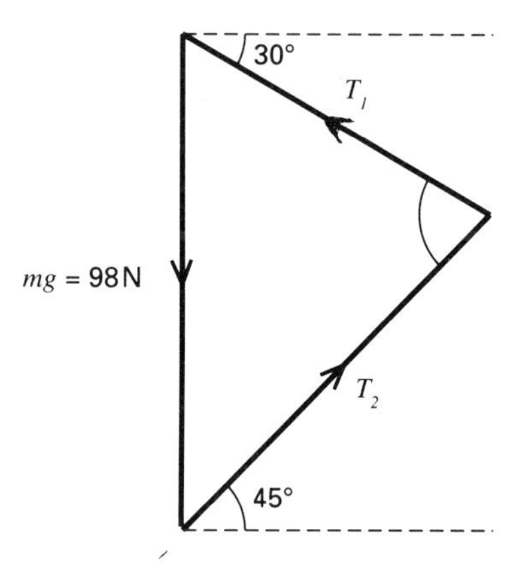

The tensions T_1 and T_2 are now measured to be 72 N and 88 N. These answers are not of course as accurate as those obtained by calculation.

Method 3 (Trigonometry)
You can however use the sine rule on the triangle above to calculate T_1 and T_2 accurately.

$$\frac{T_1}{\sin 45°} = \frac{T_2}{\sin 60°} = \frac{98}{\sin 75°}$$

$$\text{giving} \quad T_1 = \frac{98 \sin 45°}{\sin 75°} = 71.7,$$

$$T_2 = \frac{98 \sin 60°}{\sin 75°} = 87.9.$$

As before the tensions are found to be 71.7 N and 87.9 N.

EXAMPLE

A brick of mass 3 kg is at rest on a rough plane, inclined at an angle of 30° to the horizontal. Find the force F N that friction exerts on the brick, and the normal reaction R N of the plane on the brick.

Solution

The diagram shows the forces acting on the brick.

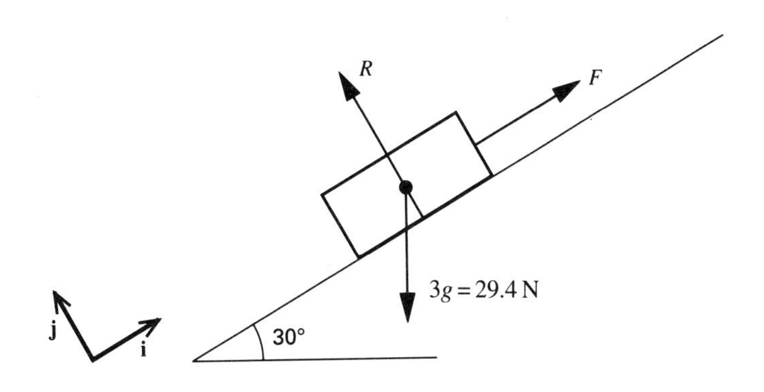

In problems like this one, we take components parallel to the plane and perpendicular to the plane and so **i** and **j** are defined as shown.

Since the brick is in equilibrium, the resultant of the three forces acting on it is zero.

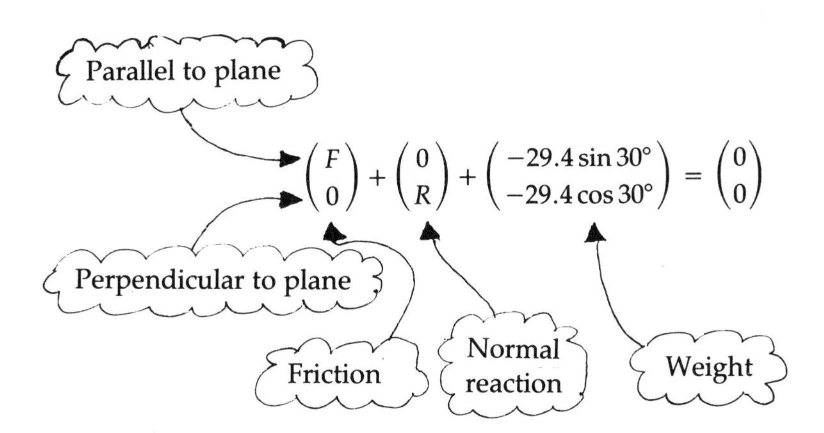

$$\begin{pmatrix} F \\ 0 \end{pmatrix} + \begin{pmatrix} 0 \\ R \end{pmatrix} + \begin{pmatrix} -29.4 \sin 30° \\ -29.4 \cos 30° \end{pmatrix} = \begin{pmatrix} 0 \\ 0 \end{pmatrix}$$

Or alternatively

$$F\mathbf{i} + R\mathbf{j} - 29.4 \sin 30°\mathbf{i} - 29.4 \cos 30°\mathbf{j} = \mathbf{0}$$

And so

Direction of **i**: $F - 29.4 \sin 30° = 0 \Rightarrow F = 14.7$

Direction of **j**: $R - 29.4 \cos 30° = 0 \Rightarrow R = 25.5$

Exercise 6B

1. The picture shows a boy, Halley, holding onto a post while his two older sisters, Michelle and Louise, try to pull him away.

Taking $\mathbf{i}$ and $\mathbf{j}$ to be unit vectors in perpendicular horizontal directions, the forces in newtons exerted by the two girls are:

Louise	$24\mathbf{i} + 18\mathbf{j}$
Michelle	$25\mathbf{i} + 60\mathbf{j}$

(i) Calculate the magnitude and direction of the force from each of the girls.

(ii) Use a scale drawing to estimate the magnitude and direction of the resultant of the forces from the two girls.

(iii) Calculate (to 3 significant figures) the magnitude and direction of the resultant from the given components in the $\mathbf{i}$ and $\mathbf{j}$ directions. Do your answers agree with those you obtained by scale drawing in part (ii)?

(iv) Do you think Halley, who is aged 7, is able to hold on to the post, or is he forced to let go?

2. The diagram shows a girder CD of mass 20 tonnes being held stationary by a crane (which is not shown). The rope from the crane (AB) is attached to a ring at B. Two ropes, BC and BD, of equal length attach the girder to B; the tensions in each of these ropes is T newtons.

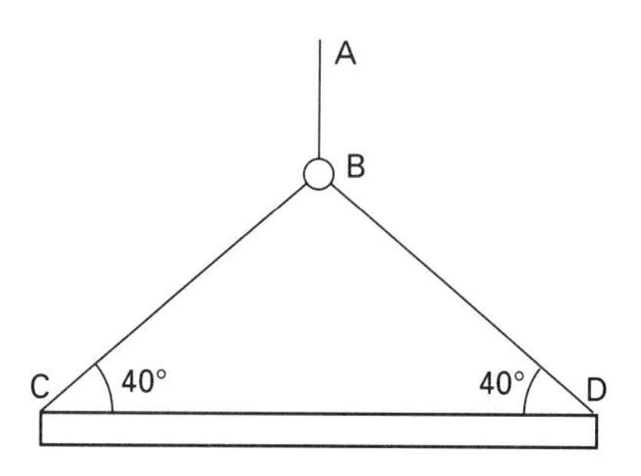

(i) Draw a diagram showing the forces acting on the girder.

(ii) Write down, in terms of T, the horizontal and vertical components of the tensions in the ropes acting at C and D.

(iii) Hence show that the tension, T, in the rope BC is 152.5 kN (to the nearest 0.1 kN).

(iv) Draw a diagram to show the three forces acting on the ring at B.

(v) Hence calculate the tension in the rope AB.

(vi) How could you have known the answer to part (v) without any calculations?

3. An angler catches a very large fish. When he tries to weigh it he finds that it is more than the 10 kg limit of his spring balance. He borrows another spring balance of exactly the same design and uses the two to weigh the fish, as shown in diagram (a). Both balances read 8 kg.

(i) What is the mass of the fish?

The angler believes the mass of the fish is a record and asks a witness to confirm it. The witness agrees with the measurements but cannot follow the calculations. He asks the angler to weigh the fish in two different positions, still using both balances. These are shown in diagrams (b) and (c). State the readings of the balances in

(ii) position (b), and (iii) position (c), assuming the spring balances themselves to have negligible mass.

(iv) Which of the three methods do you think is the best?

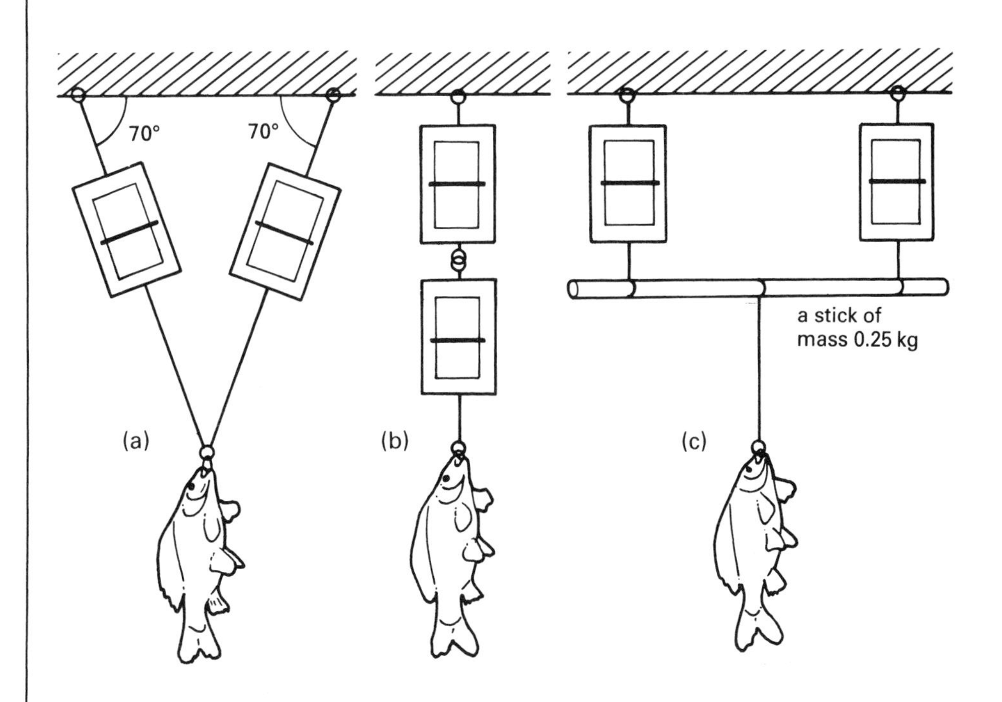

Exercise 6B continued

4. The diagram shows a device for crushing scrap cars. The light rod AB is hinged at A and raised by a cable which runs from B round a pulley at D and down to a winch at E. The vertical strut EAD is rigid and strong and AD = AB. A weight of mass 1 tonne is suspended from B by the cable BC. When the weight is correctly situated above the car the weight is released and falls onto the car.

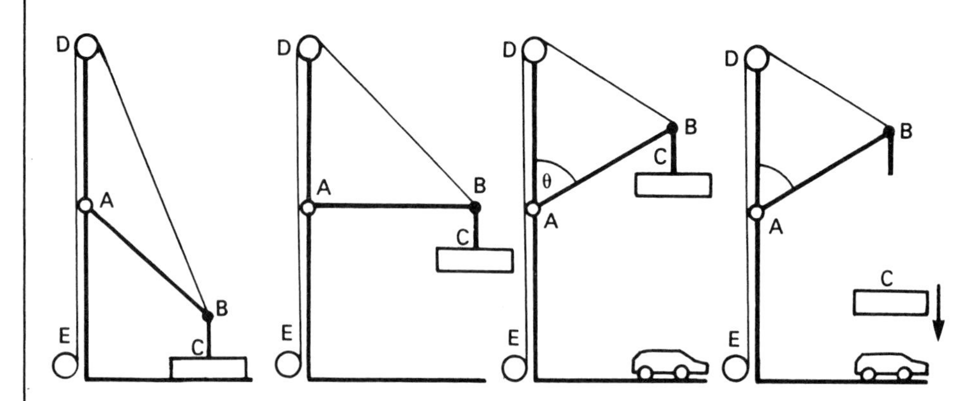

Just before the weight is released the rod AB makes angle θ with the upward vertical AD, and the weight is at rest.

(i) Draw a diagram showing the forces acting at point B in this position.

(ii) Explain why the rod AB must be in thrust and not in tension.

(iii) Draw a diagram showing the vector sum of the forces at B (i.e. the polygon of forces).

(iv) Calculate each of the three forces acting at B when
(a) $\theta = 90°$, (b) $\theta = 60°$.

5. Four wires, all of them horizontal, are attached to the top of a telegraph pole, as shown in the plan view below. The tensions in the wires are as shown.

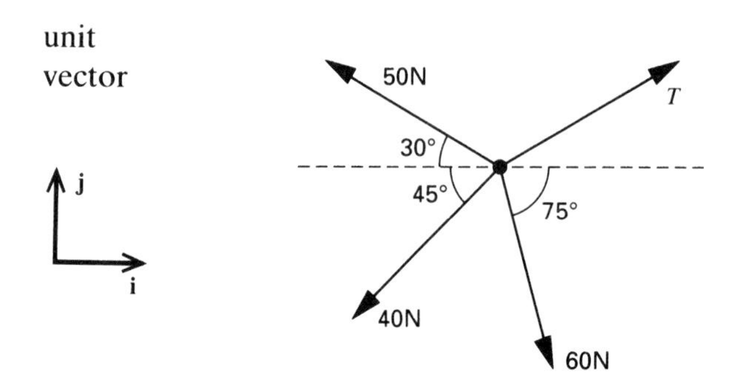

(i) Use a suitable scale drawing to estimate the magnitude and direction of the force T.

(ii) Using the unit vectors $\mathbf{i}$ and $\mathbf{j}$ shown in the diagram, show that the force of 60 N may be written as

$$15.5\mathbf{i} - 58.0\mathbf{j} \text{ (to 3 significant figures).}$$

(iii) Find T in (a) component form, (b) magnitude and direction form.

(iv) The force T is changed to $40\mathbf{i} + 35\mathbf{j}$ newtons. Show that there is now a resultant force on the pole, and find its magnitude and direction.

6. A ship is being towed by two tugs. Each tug exerts forces on the ship as indicated and there is a drag force on the ship.

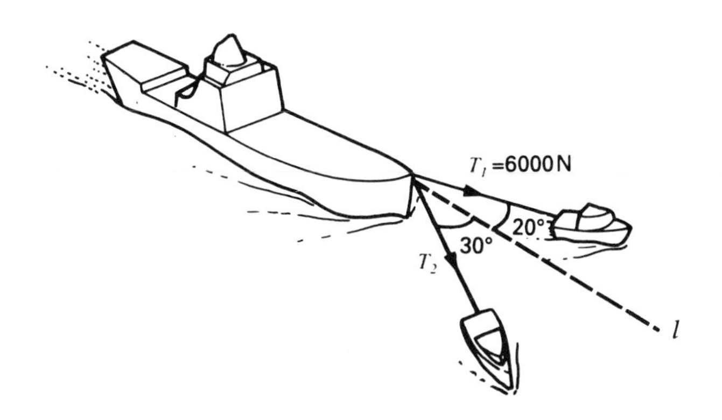

(i) Write down the components of the tensions in the towing-cables along and perpendicular to the line of motion l of the ship.

(ii) There is no resultant force perpendicular to the line l. Find T_2.

(iii) The ship is travelling with constant velocity along the line l. Find the magnitude of the drag force acting on it.

7. A boat of mass 500 kg is being winched up a beach which slopes at 10° to the horizontal. The maximum friction between the boat and the beach is 3500 N, and the rope from the boat to the winch is parallel to the slope of the beach.

The boat is on the point of moving up the beach.

(i) Draw a diagram showing all the forces acting on the boat.

(ii) Write all these forces in components parallel and perpendicular to the slope.

(iii) Find the tension in the rope.

A little later the boat is moving at a constant speed of 1 cms^{-1}.

(iv) What is the tension in the rope now?

The rope breaks.

(v) Will the boat start to slide back down the slope?

Exercise 6B continued

8. Three people are pushing a car horizontally, as shown in the diagram. The driver holds the front wheels of the car straight.

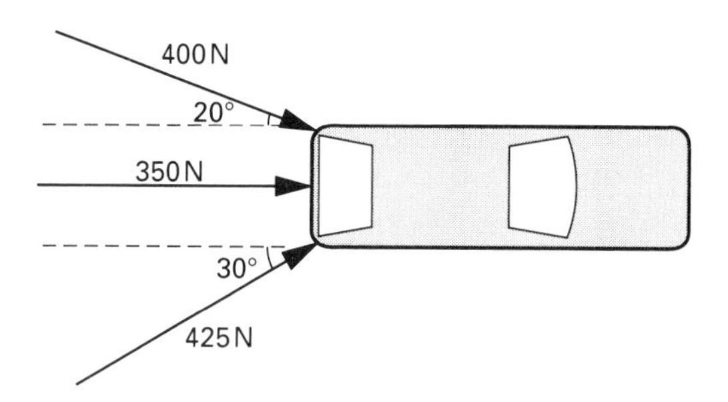

(i) Resolve the three forces into components along and perpendicular to the line of the car.
(ii) State the resultant force in each of these directions.

The resistance to motion of the car is 100 N along its line of motion and 10 000 N perpendicular to this line.
(iii) State the magnitude and direction of the overall horizontal force on the car.
(iv) Why is the resistance to motion perpendicular to the line of the car so much greater than it is along its line?

9. A skier of mass 50 kg is skiing straight down a 15° slope.
(i) Draw a diagram showing the forces acting on the skier.
(ii) Resolve these forces into components parallel and perpendicular to the slope.

The skier is travelling at constant speed.
(iii) Find the normal reaction of the slope on the skier, and the resistance force on her.

The skier later returns to the top of the slope by being pulled up it at constant speed by a rope parallel to the slope.
(iv) Assuming the resistance on the skier is the same as before, calculate the tension in the rope.

10. The diagram shows a brick of mass 5 kg on a rough inclined plane. The brick is attached to a 3 kg weight by a light string which passes over a smooth pulley. The brick is on the point of sliding up the slope.
(i) Draw a diagram showing the forces acting on the brick.
(ii) Resolve these forces into components parallel and perpendicular to the slope.
(iii) Find the force of resistance to the brick's motion.

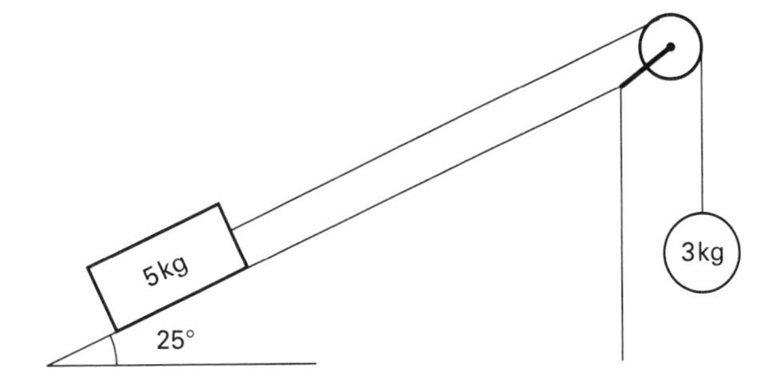

The 3 kg weight is replaced by one of mass m kg.

(iv) Find the value of m for which the brick is on the point of sliding down the slope, assuming the resistance to motion is the same as before.

11. Two husky dogs are pulling a sledge. They both exert forces of 60 N but at different angles to the line of the sledge, as shown in the diagram. The sledge is moving straight forwards.

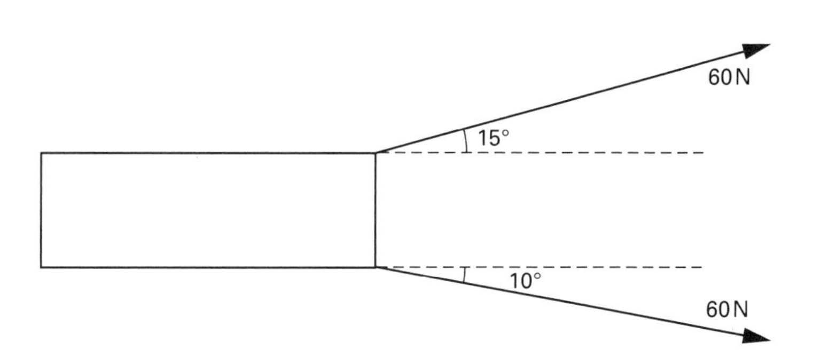

(i) Resolve the two forces into components parallel and perpendicular to the line of the sledge.

(ii) Hence find the overall forward force from the dogs, and the sideways force.

The resistance to motion is 20 N along the line of the sledge but up to 400 N perpendicular to it.

(iii) Find the magnitude and direction of the overall horizontal force on the sledge.

(iv) How much force is lost due to the dogs not pulling straight forwards?

Investigations

Pulleys

A pulley system such as a block and tackle can provide a means of lifting heavy objects easily. The diagram shows a typical pulley system.

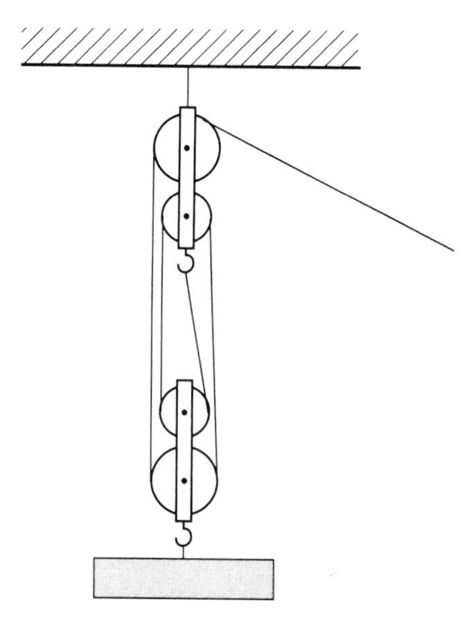

Investigate how pulley systems work and model them mathematically.

Sailing

Investigate how it is possible to sail into the wind.

KEY POINTS

- When several forces act at the same point they can be replaced by a single force called the resultant.
- If the resultant is zero the body is in equilibrium.
- If several forces acting on a point are in equilibrium, then they can be represented in magnitude and direction by a closed polygon.
- If several forces acting on a point are in equilibrium the sum of their components in any direction is zero.
- Problems involving forces in equilibrium may be approached by scale drawing or by resolving the forces into components in suitable perpendicular directions.

7 Newton's laws of motion

Nature, and Nature's Laws lay hid in Night.
God said, Let Newton be! All was Light.

Alexander Pope

Imagine that you are lucky enough to find a supermarket trolley with no tendency to veer off to one side. You position the trolley motionless in front of you and then exert a force on it by giving it a push, until it is out of your reach.

(i) What happens to the trolley while you are pushing it?

(ii) What happens to the trolley after you have stopped pushing it?

(iii) What difference does it make whether the trolley is empty or full?

The laws of motion that form the basic models in mechanics were formulated by the English mathematician, Isaac Newton. His published work, *Philosophiae Naturalis Principia Mathematica* of 1687 is perhaps one of the best examples of mathematical modelling.

HISTORICAL NOTE

Isaac Newton was born in Lincolnshire in 1642. He was not an outstanding scholar either as a schoolboy or as a university student, yet later in life he made remarkable contributions in dynamics, optics, astronomy, chemistry, music theory and theology. He became Member of Parliament for Cambridge University and later Warden of the Royal Mint. His tomb in Westminster Abbey reads 'Let mortals rejoice That there existed such and so great an Ornament to the Human Race'.

Newton's First Law of motion

Your discussion of the supermarket trolley might have led you to the following conclusions.

- Before you applied any force the trolley remained at rest.
- When you applied a force it accelerated away from you.
- When you stopped applying the force, the trolley continued to move at more-or-less constant speed – except that frictional resistance caused it to decelerate a little.

These conclusions illustrate Newton's First Law of motion:

Every body continues in a state of rest or uniform motion in a straight line unless acted on by an external force.

Newton's great insight was to realise that force is associated with a **change** in motion, rather than to motion itself. Unfortunately, many people today still associate force with velocity (or speed) rather than acceleration. For example, try out the following problem on your friends.

EXAMPLE

A girl throws a ball up into the air. The diagram shows three positions of the ball during its motion. At point A the ball is on the way up, at point B the ball is at its highest point and at point C the ball is on its way down.

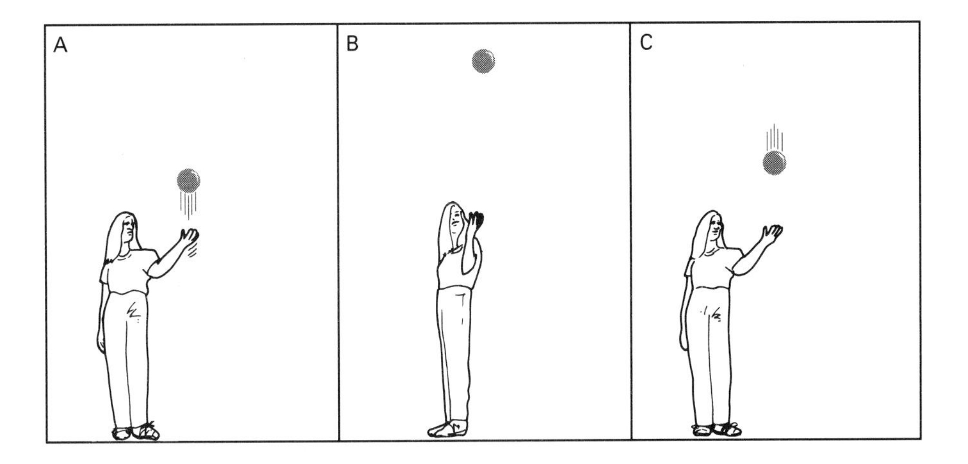

At each point draw arrows to show (i) the direction of motion of the ball, and (ii) the direction of the net force on the ball.

Solution

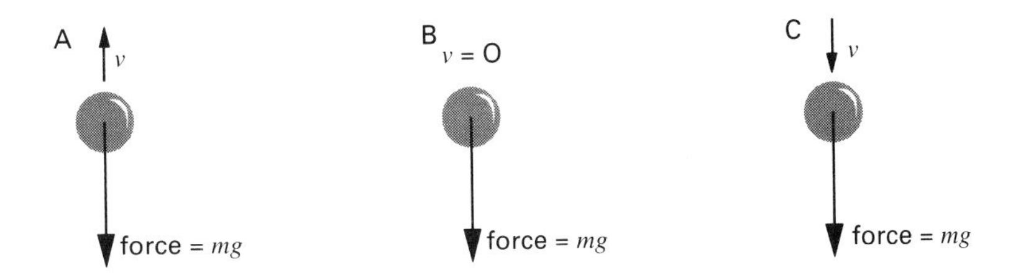

The change of motion is downwards at each instant and so the force must be downwards.

For Discussion

1. Think of three real situations which illustrate Newton's First Law of motion.
2. Imagine you are a passenger in a car. What happens to you if:
   (i) the car is suddenly accelerated from rest;
   (ii) the car drives over a hump-back bridge;
   (iii) the car drives round a sharp right-hand bend;
   (iv) the car does an emergency stop?

Exercise 7A

1. The picture shows an aeroplane flying in a horizontal circle above an airport while waiting to land.

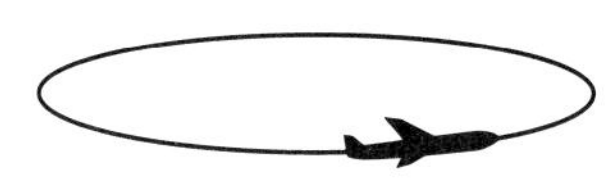

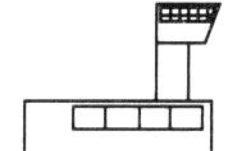

Suddenly a wheel becomes detached from the aeroplane.

(i) State in which of the directions shown below the wheel travels initially, and justify your answer.

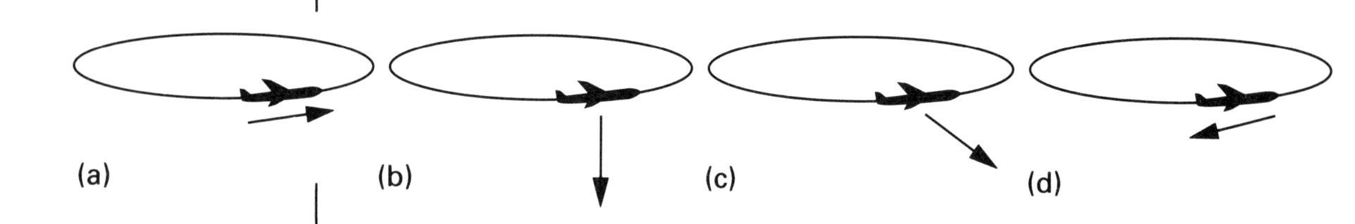

(ii) Describe the subsequent motion of the wheel. You may assume there is no wind at the time.

2. A spaceship is travelling through outer space, far from any gravitational influences, with its engines turned off.

(i) State, giving your reasons, which of the following best describes the state of the spaceship:

(a) at rest
(b) subject to constant acceleration
(c) slowing down
(d) travelling with constant velocity.

An astronaut goes outside the spaceship to carry out a repair to its hull. While there she lets go of her spanner by mistake.

(ii) Describe what happens to the spanner.

She later breaks a screwdriver of mass 0.25 kg and throws it away.

(iii) How far away can she throw the screwdriver?

3. A diver is at rest near the bottom of the sea.

(i) Draw a diagram showing the forces acting on him.

He has a number of weights attached to his belt. He detaches one of them and lets it go.

(ii) Draw another diagram showing the forces that are now acting on the diver and also those acting on the detached weight.

(iii) Describe the subsequent motions of both the diver and the weight.

4. A parachutist jumps from an aeroplane and freefalls for 20 seconds before opening her parachute. By the time she reaches the ground she is travelling at a steady speed of $5\ \mathrm{ms}^{-1}$. The combined mass of the parachutist and her parachute is 60 kg. It takes her 150 seconds to reach the ground.

(i) What is the magnitude and direction of the air resistance acting on the parachute just before she reaches the ground?

(ii) What can you say about the magnitude and direction of the air resistance (a) 10 seconds, and (b) 22 seconds after she jumped out of the aeroplane?

5. A car of mass 800 kg, when moving at a speed of 30 mph, is subject to a total resistance force of 200 N. Describe its motion at this speed when

(i) the driving force is 200 N,

(ii) the driving force is 300 N,

(iii) it has just run of out petrol.

6. A helicopter, taking off vertically, is gaining height at a constant rate. The lift force from the helicopter's rotors is 24 500 N.

(i) What is the mass of the helicopter?

(ii) What would happen to the helicopter if, unnoticed, one of its passengers fell out?

Another helicopter is unable to lift off the ground despite the fact that its rotor blades are also giving an upthrust of 24 500 N.

(iii) What can you deduce about this helicopter?

Exercise 7A continued

7. Joanna has pushed her loaded supermarket trolley to the top of a 5° slope when her children start fighting. While she is sorting them out she inadvertently lets go of her trolley and it starts rolling back down the slope. After travelling a short distance it runs at a steady speed of 1.5 ms^{-1}. The combined mass of the trolley and its contents is 30 kg.

(i) Calculate the total resistance force on the trolley when it is travelling at that speed.

(ii) When Joanna has finished dealing with her children the trolley has a 10 m lead on her. Estimate how much further it will travel before she catches it up.

8. A cyclist freewheels at a constant speed of 6 ms^{-1} down a slope which makes an angle of 2° to the horizontal. The mass of the cyclist is 62 kg and that of the bicycle is 12 kg.

(i) What is the total force of resistance on the cyclist and his bicycle?

If the cyclist adopts a crouched position when freewheeling down the same hill he finds that he reaches a steady speed of 8 ms^{-1}.

(ii) Explain very carefully what you may conclude from this observation.

Newton's Second Law of motion

Newton's First Law tells us that an object will not change its motion unless it experiences a net force. Newton's Second Law provides a link between the force and the change in motion:

If a force F acts on a mass m, then it produces an acceleration a, where

$$\mathbf{F} = m\mathbf{a}.$$

This is called the equation of motion of the object.

Whenever a net force acts on a body an acceleration is produced. Remember that acceleration is a vector quantity. The direction of the acceleration is the same as that of the force. Its magnitude is proportional to that of the force but inversely proportional to the mass of the object. This is illustrated by the fact that when you push a loaded supermarket trolley its acceleration is less than when you push an empty one, as you probably decided when considering the example at the start of the chapter.

EXAMPLE

A lift and its passengers have a total mass of 400 kg. Find the tension in the cable supporting the lift if

(i) the lift is at rest,

(ii) the lift is moving at constant speed,

(iii) the lift is accelerating upwards at 0.8 ms^{-2},

(iv) the lift is accelerating downwards at 0.6 ms^{-2}.

Solution:
Before starting calculations you must define a direction as positive. In this example the upward direction is chosen to be positive.

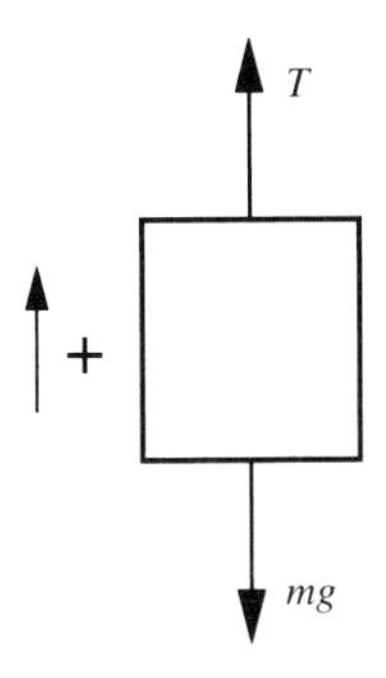

(i) **At rest**
As the lift is at rest the forces must be in equilibrium.

$$T - mg = 0$$
$$T - 400 \times 9.8 = 0$$
$$T = 3920 \text{ N}.$$

(ii) **Ascending at constant speed**
Here again the forces on the lift must be in equilibrium, because it is moving at a constant speed, so $T = 3920$ N.

(iii) **Accelerating upwards**
The resultant upward force on the lift is $T - mg$ so the equation of motion is

$$T - mg = ma$$

which in this case gives

$$T - 400 \times 9.8 = 400 \times 0.8$$
$$T - 3920 = 320$$
$$T = 4240 \text{ N}.$$

(iv) **Accelerating downwards**
The equation of motion is

$$T - mg = ma$$

A downward acceleration of 0.6 ms^{-2} is an upward acceleration of -0.6 ms^{-2}.

In this case, we have

$$T - 400 \times 9.8 = 400 \times (-0.6)$$
$$T - 3920 = -240$$
$$T = 3680 \text{ N}.$$

EXAMPLE

Two forces, $\mathbf{F}_1 = 4\mathbf{i} + 12\mathbf{j}$, and $\mathbf{F}_2 = 3\mathbf{i} - 2\mathbf{j}$ (in newtons) act on a particle of mass 2 kg. Find the acceleration of the particle.

Solution:

First we find the resultant, $\mathbf{F}$, of the two forces.

$$\begin{aligned}\mathbf{F} &= \mathbf{F}_1 + \mathbf{F}_2 \\ &= 4\mathbf{i} + 12\mathbf{j} + 3\mathbf{i} - 2\mathbf{j} \\ &= 7\mathbf{i} + 10\mathbf{j}\end{aligned}$$

Newton's Second Law states:

$$\mathbf{F} = m\mathbf{a}$$

So the equation of motion is

$$\begin{aligned}7\mathbf{i} + 10\mathbf{j} &= 2\mathbf{a} \\ \mathbf{a} &= 3.5\mathbf{i} + 5\mathbf{j}\end{aligned}$$

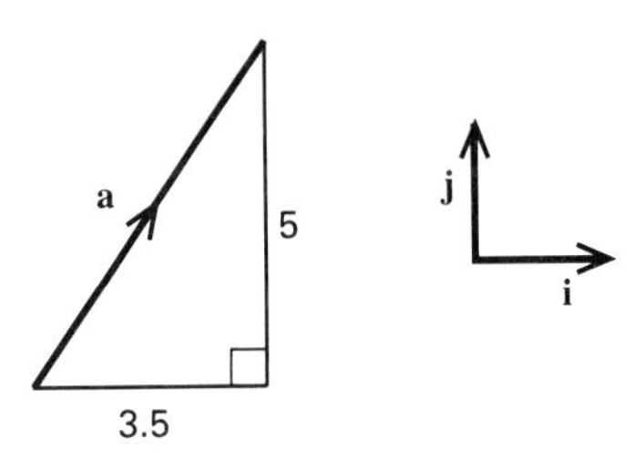

The acceleration has magnitude $\sqrt{3.5^2 + 5^2} = 6.10 \text{ ms}^{-2}$

The direction is $\arctan\left(\frac{5}{3.5}\right)$ to the $\mathbf{i}$ direction, i.e. 55° to the $\mathbf{i}$ direction.

Exercise 7B

1. Calculate the force in newtons acting in the following situations.

(i) A car of mass 400 kg has acceleration 2 ms^{-2}.

(ii) A blue whale of mass 177 tonnes has acceleration ½ ms^{-2}.

(iii) A pygmy mouse of mass 7.5 g has acceleration of 3 ms^{-2}.

(iv) A freight train of mass 42 000 tonnes brakes with deceleration of 0.02 ms^{-2}.

(v) A bacterium of mass 2×10^{-16} g has acceleration 0.4 mms^{-2}.
(vi) A woman of mass 56 kg falling off a high building has acceleration 9.8 ms^{-2}.
(vii) A jumping flea of mass 0.05 mg accelerates at 1750 ms^{-2} during take-off.
(viii) A galaxy of mass 10^{42} kg has acceleration 10^{-12} ms^{-2}.

2. A force of 100 N is applied to a body. Find the mass of the body when its acceleration is
(i) 0.5 ms^{-2}, (ii) 2 ms^{-2},
(iii) 0.01 ms^{-2}, (iv) $10g$.

3. A man pushes a car of mass 400 kg on level ground with a force of 200 newtons. The car is initially at rest and the man maintains this force until the car reaches a speed of 5 ms^{-1}. Ignoring any resistance forces, find
(i) the acceleration of the car,
(ii) the distance the car travels while the man is pushing.

4. The engine of a car of mass 1.2 tonnes can produce a driving force of 2000 N. Ignoring any resistance forces, find
(i) the car's resulting acceleration,
(ii) the time taken for the car to go from rest to 27 ms^{-1} (about 60 mph).

5. A top sprinter of mass 65 kg reaches a speed of 10 ms^{-1} in 2 s.
(i) Calculate the force required to produce this acceleration, assuming it is uniform.
(ii) Compare this to the force exerted by a weightlifter holding a mass of 180 kg above the ground.

6. A supertanker of mass 500 000 tonnes is travelling at a speed of 10 ms^{-1} when its engines fail. It then takes half an hour for the supertanker to stop.
(i) Find the force of resistance, assuming it to be constant, acting on the supertanker.

When the engines have been repaired it takes the supertanker 10 minutes to return to full speed of 10 ms^{-1}.
(ii) Find the driving force produced by the engines, assuming this also to be constant.

7. A helicopter of mass 1000 kg is taking off vertically.
(i) Draw a labelled diagram showing the forces on the helicopter as it lifts off, and the direction of its acceleration.
(ii) Its initial upward acceleration is 1.5 ms^{-2}. Find the upward force its rotors exert. Ignore the effects of air resistance.

Exercise 7B continued

8. The forces $\mathbf{F}_1 = 4\mathbf{i} - 5\mathbf{j}$ and $\mathbf{F}_2 = 2\mathbf{i} + \mathbf{j}$, in newtons, act on a particle of mass 4 kg.
(i) Find the acceleration of the particle in component form.
(ii) Find the magnitude of the particle's acceleration.

9. Two forces $\mathbf{P}_1$ and $\mathbf{P}_2$ act on a particle of mass 2 kg giving it an acceleration $5\mathbf{i} + 5\mathbf{j}$ (in ms^{-2}).
(i) If $\mathbf{P}_1 = 6\mathbf{i} - \mathbf{j}$ (in newtons), find $\mathbf{P}_2$.
(ii) If instead $\mathbf{P}_1$ and $\mathbf{P}_2$ both act in the same direction but $\mathbf{P}_1$ is four times as big as $\mathbf{P}_2$, find both forces.

10. An ice skater of mass 65 kg is initially moving with speed 2 ms^{-1} and glides to a halt over a distance of 10 m. Assuming that the force of resistance is constant, find
(i) the size of the resistance force,
(ii) the distance he would travel gliding to rest from an initial speed of 6 ms^{-1},
(iii) the force he would need to apply to maintain a steady speed of 10 ms^{-1}.

11. A crate of mass 30 kg is being pulled up a smooth slope inclined at 30° to the horizontal by a rope which is parallel to the slope. The crate has acceleration 0.75 ms^{-2}.
(i) Draw a diagram showing the forces acting on the crate, and the direction of its acceleration.
(ii) Resolve the forces in directions parallel and perpendicular to the slope.
(iii) Find the tension in the rope.
(iv) The rope suddenly snaps. What happens to the crate?

12. A cyclist of mass 60 kg rides a bicycle of mass 7 kg. The greatest forward force that she can produce is 200 N but she is subject to air resistance and friction totalling 50 N.
(i) Draw a diagram showing the forces acting on the cyclist when she is going uphill.
(ii) What is the angle of the steepest slope that she can ascend?
(iii) The cyclist comes to a long steady uphill slope of angle 20°. Describe what happens to her.

13. A particle of mass 0.5 kg is acted on by a force, in newtons, $\mathbf{F} = t^2\mathbf{i} + 2t\mathbf{j}$. The particle is initially at rest at the origin, and t is measured in seconds.
(i) Find the acceleration of the particle at time t.
(ii) Find the velocity of the particle at time t.
(iii) Find the position vector of the particle at time t.
(iv) Give all the information you can about the particle at time 2 seconds.

Newton's Third Law of motion

For Discussion

The photograph shows two ice skaters pushing against each other.

What can you say about the force they exert on each other, and about their acceleration and subsequent motion across the ice?

Newton's Third Law of motion describes the forces of interaction when two objects are in contact. If two objects A and B exert forces on each other then the force exerted on A by B is equal in magnitude and opposite in direction to the force exerted on B by A. This is often stated in the form:
To each force there is an equal and opposite reaction.

EXAMPLE

Two skaters are facing each other on an ice rink. The lighter skater, of mass 50 kg, pushes the heavier one, of mass 65 kg, exerting a force of 25 N on him for the 0.5 seconds that they are in contact. They experience negligible resistance to motion. Find their velocities when they lose contact.

Solution
Define the positive direction to be to the right. Since the heavier skater exerts a force equal in magnitude, but opposite in direction, on the lighter skater the arrows on the diagram point in opposite directions.

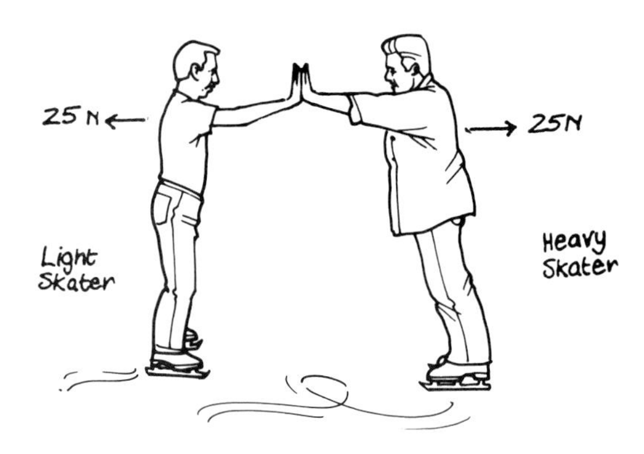

Light skater		**Heavy skater**	
Net force $= -25$ N		Net force $= 25$ N	
Acceleration	$= -25/50\ \text{ms}^{-2}$	Acceleration	$= 25/65\ \text{ms}^{-2}$
	$= -0.5\ \text{ms}^{-2}$		$= 0.38\ \text{ms}^{-2}$
Velocity	$= -0.5t\ \text{ms}^{-1}$	Velocity	$= 0.38t\ \text{ms}^{-1}$

The skaters lose contact and the force stops acting after 0.5 s. After this their velocities are:

Heavy skater	0.38×0.5	$= 0.19\ \text{ms}^{-1}$
Light skater	-0.5×0.5	$= -0.25\ \text{ms}^{-1}$

Exercise 7C

In this exercise you should take the value of g to be $10\ \text{ms}^{-2}$.

1. Two railway trucks of different masses roll towards each other with different speeds. When they collide, each exerts a force on the other through its buffers. Are these statements true or false?
   (i) The heavier truck exerts the greater force.
   (ii) The faster truck exerts the greater force.
   (iii) They both exert the same force.
   (iv) The truck with the stronger buffer exerts the greater force.

2. The picture shows a man training with a chest expander. As you can see he has pulled it much further with his left hand than he has with his right hand. Which of these statements about the magnitude of the forces are true and which are false?

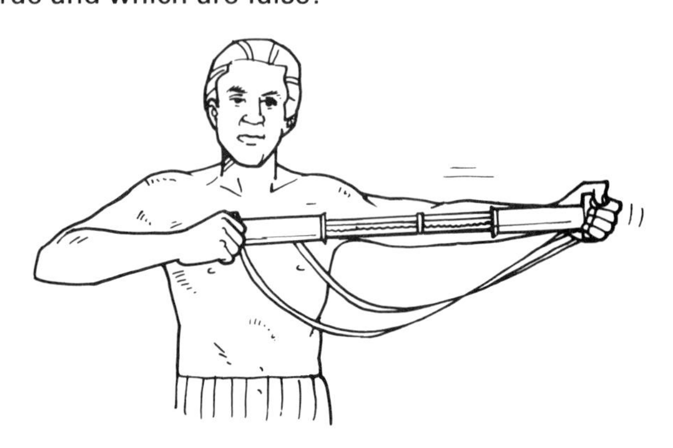

(i) His left hand is exerting a greater force than his right hand.
(ii) His right hand is exerting a greater force than his left hand.
(iii) The springs must be weaker on the left side than they are on the right side of the chest expander.
(iv) Both the man's hands are exerting equal forces on the chest expander.

3. A girl of mass 54 kg dives off the stern of a rowing boat of mass 300 kg. Just before she loses contact with the boat she is exerting a horizontal force of 100 N on it.
(i) What force is the boat exerting on the girl at the moment described?
(ii) Describe the motion of both the girl and the boat at that moment.
(iii) What happens to the boat after the girl has left it?

4. The picture shows a boy retrieving a toy of mass ½ kg from on top of a cupboard. The boy holds the toy in one hand and jumps off the stool. What force does the toy exert on his hand
(i) before he jumps down?
(ii) when he is in mid-air?
(iii) some time after he has landed?

5. A man who weighs 720 N can reach to a height of 2.1 m. He is doing some repairs on a shed, the ceiling of which is 2.05 m above its floor. In each of these situations (a) draw a force diagram, (b) state whether the named force can in theory exceed 720 N.
(i) He is pushing upwards on the ceiling with force U.
(ii) He is pulling downwards on the ceiling with force D.
(iii) He is pulling upwards on a nail in the floor with force F.
(iv) He is pushing downwards on the floor with force T.

Exercise 7C continued

6. A bullet of mass 50 g accelerates at a constant rate of 10^4 ms^{-2} while in the barrel of the gun, which is 80 cm long. The mass of the gun is 20 kg.

(i) Find the force acting on the bullet.

(ii) Find the force acting on the gun.

(iii) Find the speed of the bullet when it leaves the barrel of the gun.

(iv) Find the speed of the barrel of the gun as the bullet leaves it, assuming it is not restrained.

(v) Are your answers to parts (iii) and (iv) altered if the acceleration is not constant and the value given is the average acceleration?

7. The diagram shows a train, consisting of an engine of mass 50 000 kg pulling two trucks, A and B, each of mass 10 000 kg. The force of resistance on the engine is 2000 N and that on each of the trucks 200 N. The train is travelling at constant speed.

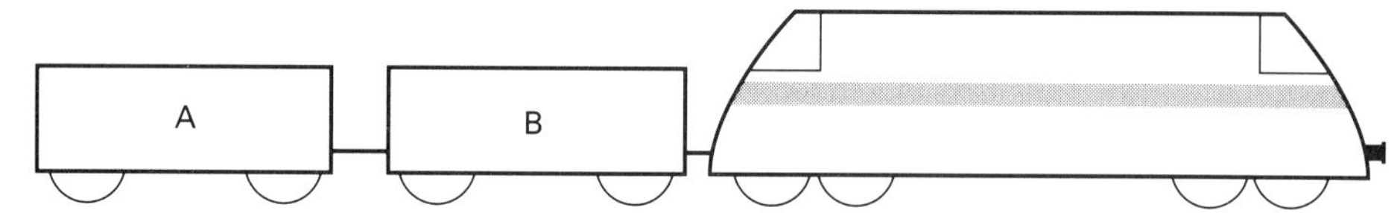

(i) By considering the equilibrium of the train as a whole, find the driving force provided by the engine.

The coupling connecting truck A to the engine exerts a force T_1 on the engine.

(ii) By considering the equilibrium of the engine alone, find T_1.

(iii) What force does the coupling from the engine exert on truck B?

The coupling connecting truck B to truck A exerts a force T_2 on truck A.

(iv) By considering the equilibrium of truck A alone, find T_2.

(v) Show that truck B is also in equilibrium.

8. The diagram shows a lift of mass 450 kg containing a single passenger whose mass is 50 kg. The lift is stationary.

Exercise 7C continued

(i) Make a simple sketch to show each of the following forces:

T the tension in the cable,

R_P the reaction of the lift on the passenger,

R_L the reaction of the passenger on the lift,

Mg the weight of the passenger.

(ii) Calculate T, R_P and R_L.

The lift then accelerates upwards at 0.8ms^{-2}.

(iii) Find the new values of T, R_P and R_L.

Problem solving with Newton's Laws

You are now in a position to solve many problems in mechanics, as you will see in the following examples and in Exercise 7D.

EXAMPLE

A skier is being pulled up a 20° dry ski slope by a rope which makes an angle of 30° with the horizontal. The mass of the skier is 75 kg and the tension in the rope is 300 N. Initially the skier is at rest at the bottom of the slope. The slope is smooth.

Find the skier's speed after 5 seconds, and find the distance he has travelled by that time.

Solution

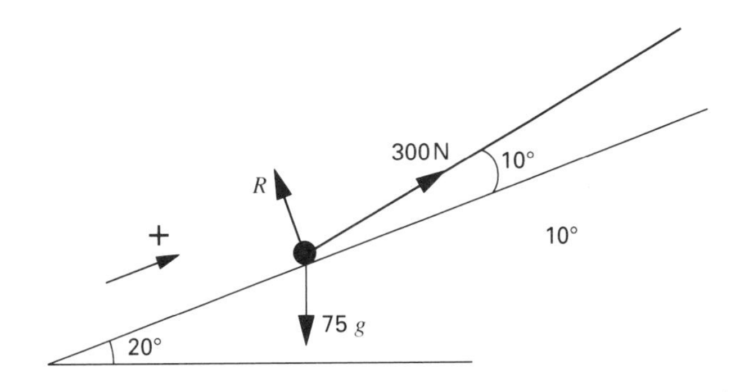

In the diagram the skier is modelled as a particle.
Since the skier moves parallel to the slope, we consider motion in that direction.

$$F = ma$$

$$300\cos 10° - 75g\cos 70° = 75a$$

Taking g as 9.8 ms^{-2},

$$a = \frac{44.06}{75} = 0.59 \text{ ms}^{-2} \text{ (to 2 decimal places).}$$

This is a constant acceleration, so we use the constant acceleration formulae, with $u = 0$, $a = 0.59$, $t = 5$.

Speed: $v = u + at = 0 + 0.59 \times 5$

$= 2.94 \text{ ms}^{-1}$ (to 2 decimal places).

Distance travelled:

$$s = ut + \tfrac{1}{2}at^2 = 0 + \tfrac{1}{2} \times 0.59 \times 5^2$$

$$= 7.34 \text{ m (to 2 decimal places).}$$

EXAMPLE

A brick of mass 3 kg lies on a smooth horizontal table. It is connected to a stone of mass 5 kg by a light inextensible string which passes over a smooth pulley fixed at the edge of the table so that the stone hangs vertically. Initially the system is held at rest.

A horizontal force of magnitude 85 N is then applied to the brick in a direction such that the stone is raised.

(i) Draw a diagram showing all the forces acting on the brick and the stone.

(ii) Write down the equations of motion of (a) the brick (b) the stone.

(iii) Find the acceleration of the brick and the stone.

(iv) Find the tension in the string.

(v) What happens if you write down and solve the equation of motion for the external forces acting in the direction of motion for the system as a whole?

Solution

(i) T is the tension in the string, R the normal reaction of the table on the brick, both in newtons.

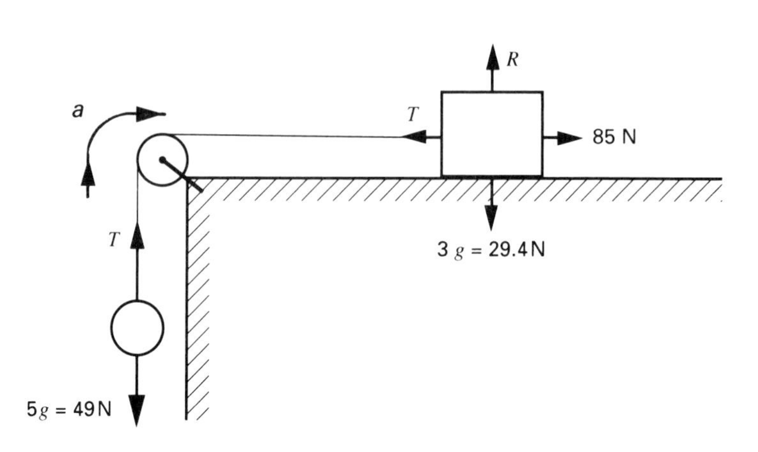

(ii) Since the string is inextensible, both the brick and the stone must have the same acceleration, a in ms^{-2}. Using Newton's Second Law, the equations of motion are as follows.

(a) The brick moves horizontally from left to right with acceleration a given by

$$85 - T = 3a \qquad (1)$$

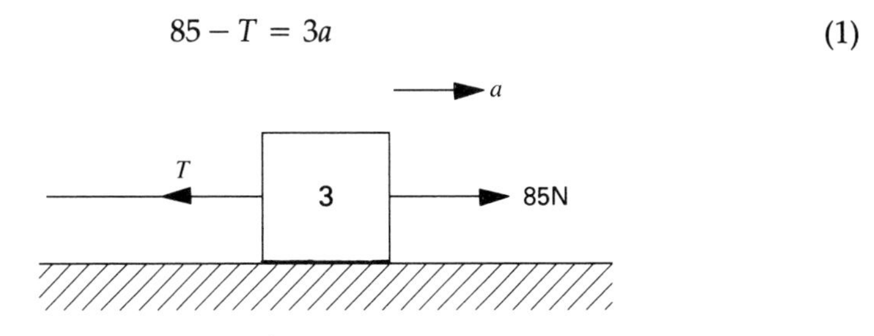

(b) The stone moves vertically upwards with the same acceleration a, given by

$$T - 49 = 5a \qquad (2)$$

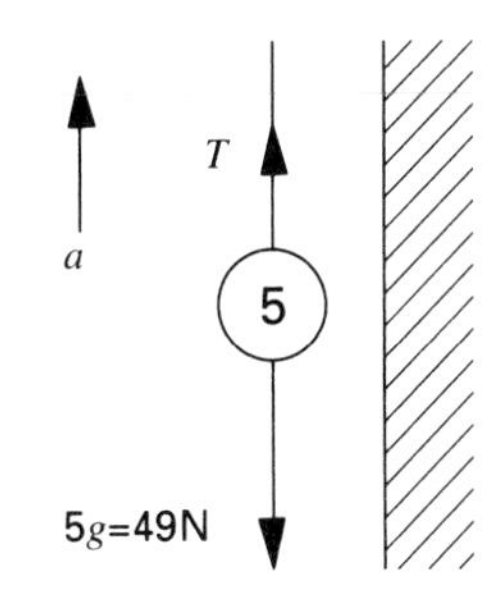

(iii) Adding equations (1) and (2) eliminates the tension T

$$85 - T = 3a \qquad (1)$$

$$T - 49 = 5a \qquad (2)$$

$$36 = 8a$$

$$a = {}^{36}/_{8}$$

The acceleration is $4.5\ ms^{-2}$.

(iv) Substituting $a = 4.5$ in equation (2) gives

$$T - 49 = 5 \times 4.5$$

$$T = 22.5 + 49 = 71.5$$

The tension is 71.5 N.

NOTE *You can check these results as follows.*

The brick: $85 - 71.5 = 13.5 = 3 \times 4.5$

The stone: $71.5 - 49 = 22.5 = 5 \times 4.5$

(v) The external forces in the direction of motion are the applied force of 85 N and the weight 49 N. The tension in the string is an internal force. The mass of the whole system is that of the brick and the stone together, $5 + 3 = 8$ kg.
The equation of motion is thus

$$85 - 49 = 8a$$

This simplifies to $36 = 8a \Rightarrow a = 4.5$, as before.

The same answer is obtained for the acceleration.

NOTE

You will see that in this case it was possible to find the acceleration by treating the system as a whole. This approach may be extended to many situations which are modelled as connected particles but you must be careful about using it. The equation F = ma *is being applied in the overall direction of motion and so not as a vector equation as usual. In addition you must be certain that you are dealing with the whole system and not part of it, which may itself be accelerating. It is usually safer to split the system into parts (as in (ii), (iii), (iv) of the example above); this avoids the need to decide which forces are external and which are internal.*

EXAMPLE

A cable-car has mass 500 kg and travels along a cable inclined at 30° to the horizontal. It is supported by two stays making angles of 30° and 50° to the cable, as shown. For the first 10 seconds of its motion the cable-car has acceleration 0.1 ms^{-2} parallel to the cable. Find the tension in each stay.

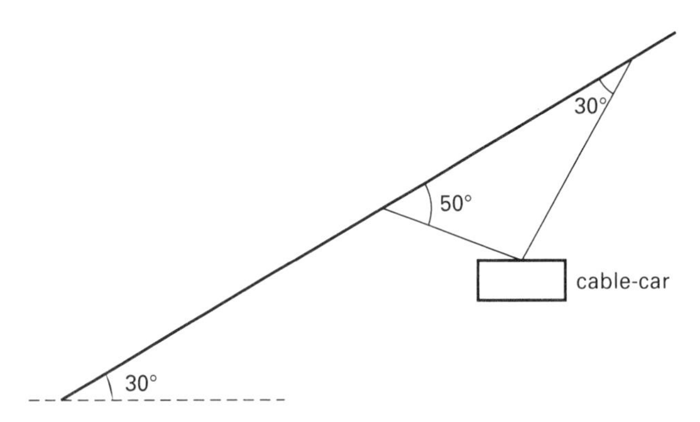

Solution

Assume the cable to be straight and the stays to be inextensible.

Take the origin to be the starting position of the cable-car and take axes parallel and perpendicular to the line of the cable.
There are 3 forces acting on the cable-car, the tensions T_1 and T_2 in its two stays and its weight, $500g$ N. These are resolved into components parallel to the axes as shown in the diagram.

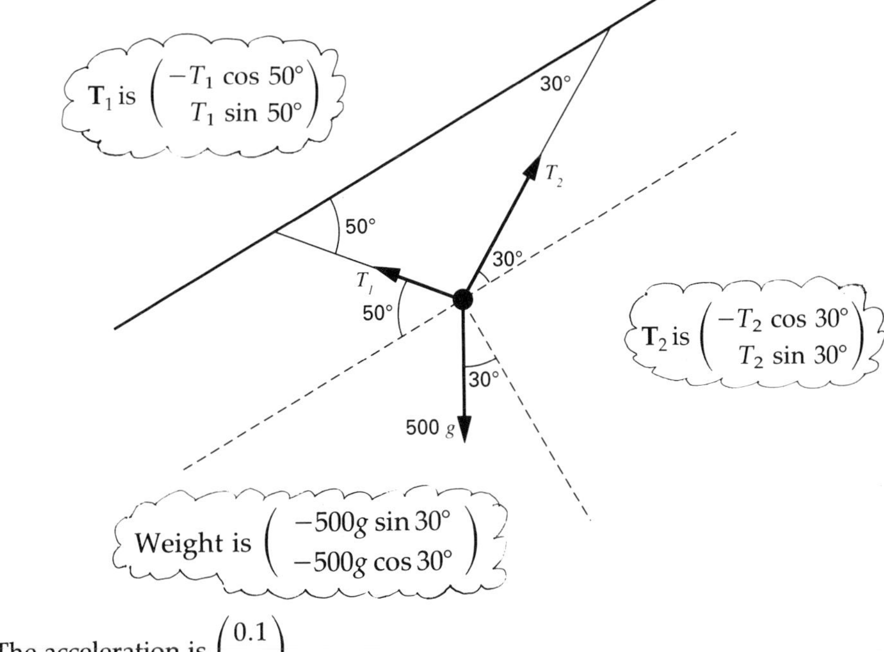

The acceleration is $\begin{pmatrix} 0.1 \\ 0 \end{pmatrix}$

The equation of motion is

Parallel to the cable

$$\begin{pmatrix} -T_1 \cos 50° \\ T_1 \sin 50° \end{pmatrix} + \begin{pmatrix} T_2 \cos 30° \\ T_2 \sin 30° \end{pmatrix} + \begin{pmatrix} -500g \sin 30° \\ -500g \cos 30° \end{pmatrix} = 500 \times \begin{pmatrix} 0.1 \\ 0 \end{pmatrix}$$

Perpendicular to the cable

Using $g = 9.8\ \text{ms}^{-2}$, this simplifies to two simultaneous equations:

Parallel to cable: $\quad -0.6428T_1 + 0.8660T_2 = 2500$

Perpendicular to cable: $\quad 0.7660T_1 + 0.5T_2 = 4244$

Solving these gives

$$T_1 = 2462\ \text{N}$$
$$T_2 = 4716\ \text{N}$$

Exercise 7D

1. A crane is used to lift a hopper full of cement to a height of 20 m on a building site. The hopper has mass 200 kg and the cement 500 kg. Initially the hopper accelerates upwards at 0.05 ms^{-2}, then it travels at constant speed for some time before decelerating at 0.1 ms^{-2} until it is at rest. The hopper is then emptied.
   (i) Find the tension in the crane's cable during each of the three phases of the motion, and after emptying.

Exercise 7D continued

The cable's maximum safe load is 10 000 N.

(ii) What is the greatest mass of cement that can safely be transported in the same manner?

The cable is in fact faulty and on a later occasion breaks without the hopper leaving the ground. On that occasion the hopper is loaded with 720 kg of cement.

(iii) What can you say about the strength of the cable?

2. A man of mass 70 kg is standing in a lift which has an upward acceleration a ms^{-2}.

(i) Draw a diagram showing the man's weight, the force, R, that the lift floor exerts on him and the direction of his acceleration.

(ii) Taking g to be 10 ms^{-2}, find the value of a when $R = 770$ N.

The graph below shows the value of R from the time ($t = 0$ s) when the man steps into the lift to the time ($t = 12$ s) when he steps out.

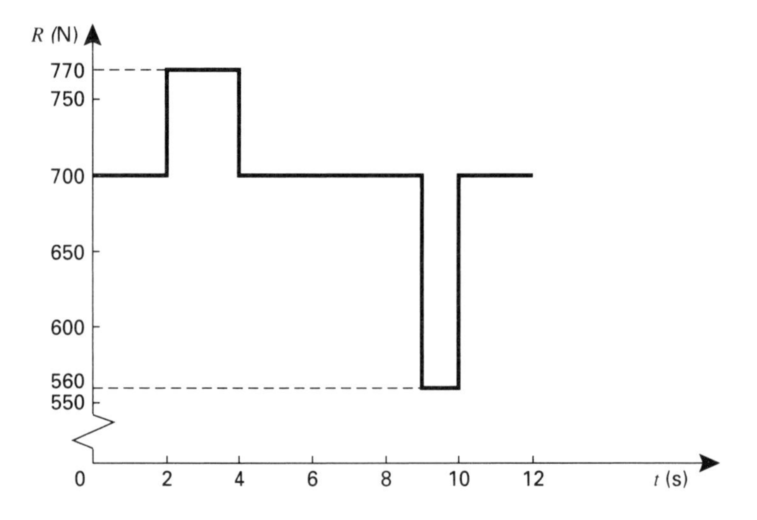

(iii) Explain what is happening in each section of the journey.

(iv) Draw the corresponding speed–time graph.

(v) To what height does the man ascend?

3. In this question you should take g to be 10 ms^{-2}. The diagram shows a block of mass 5 kg lying on a smooth table. It is attached to blocks of mass 3 kg and 2 kg by strings which pass over smooth pulleys. The tensions in the strings are T_1 and T_2 as shown, and the blocks have acceleration a.

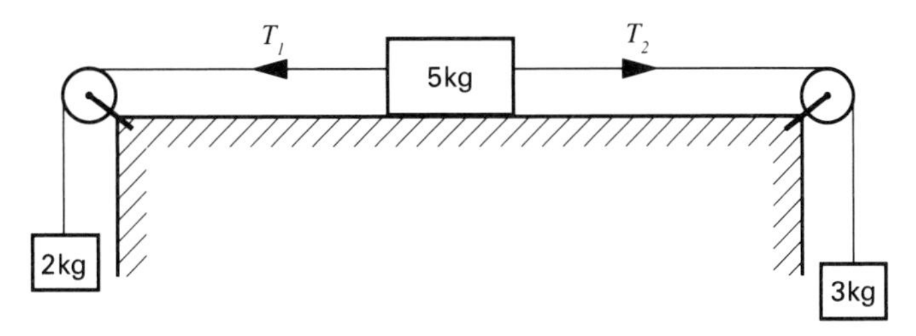

(i) Copy the diagram and mark in all the forces acting on the blocks.
(ii) Write down the equation of motion for each of the blocks.
(iii) Hence find the values of a, T_1 and T_2.

In practice the value of a is found to be 0.5 ms^{-2}. This is because the table is not truly smooth.
(iv) What frictional force would produce this result?

4. A builder is demolishing the chimney of a house and slides the old bricks down to the ground on a straight chute 10 m long inclined at 42° to the horizontal. Each brick has mass 3 kg.
(i) Draw a diagram showing the forces acting on a brick as it slides down the chute, assuming the chute to have a flat cross section and a smooth surface.
(ii) Find the acceleration of the brick.
(iii) Find the time the brick takes to reach the ground.

In fact the chute is not smooth and the brick takes 3 seconds to reach the ground.
(iv) Find the frictional force acting on the brick, assuming it to be constant.

5. The diagram shows a goods train consisting of an engine of mass 40 tonnes and two trucks of 20 tonnes each. The engine is producing a driving force of 5×10^4 N, causing the train to accelerate. The ground is level and resistance forces may be neglected.

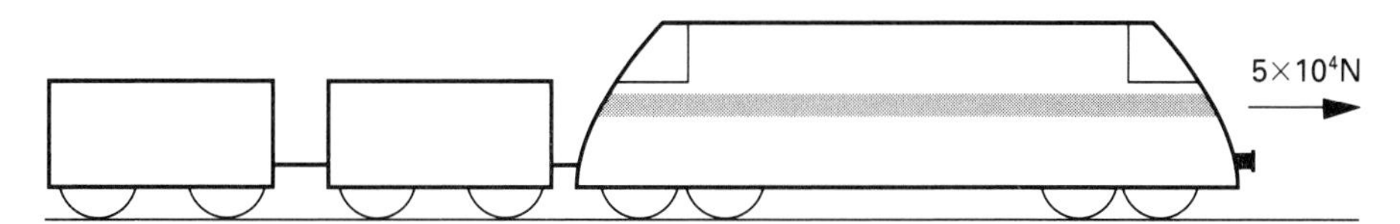

(i) Find the acceleration of the train.
(ii) Draw a diagram to show the forces acting on the truck next to the engine.
(iii) Find the tensions in each of the two couplings.

The brakes on the first truck are faulty and suddenly engage, causing a resistance of 10^4 N.
(iv) What effect does this have on the tension in the coupling to the last truck?

MEI

Exercise 7D continued

6. The police estimate that for good road conditions the frictional force, F, on a skidding vehicle of mass m is given by $F = 0.8\,mg$. A car of mass 450 kg skids to a halt narrowly missing a child. The police measure the skid marks and find they are 12.0 m long.

(i) Calculate the deceleration of the car when it was skidding to a halt.

The child's mother says the car was travelling well over the speed limit but the driver of the car says she was travelling at 30 mph and the child ran out in front of her.

(ii) Calculate the speed of the car when it started to skid. Who was telling the truth?

7. A lift in a mineshaft takes exactly one minute to descend 500 m. It starts from rest, accelerates uniformly for 12.5 seconds to a constant speed which it maintains for some time, and then decelerates uniformly to stop at the bottom of the shaft. The mass of the lift is 5 tonnes and on the day in question it is carrying 12 miners whose average mass is 80 kg.

(i) Sketch the speed–time graph of the lift.

During the first stage of the motion the tension in the cable is 53 640 N.

(ii) Find the acceleration of the lift during this stage.

(iii) Find the length of time for which the lift is travelling at constant speed, and find the final deceleration.

(iv) What is the maximum value of the tension in the cable?

(v) Just before the lift stops one miner experiences an upthrust of 1002 N from the floor of the lift. What is the mass of the miner?

8. The diagram shows a girl pulling a sledge at steady speed across level snow-covered ground using a rope which makes an angle of 30° to the horizontal. The mass of the sledge is 8 kg and there is a resistance force of 10 N.

(i) Draw a diagram showing the forces acting on the sledge.

(ii) Find the magnitude of the tension in the rope.

The girl comes to an area of ice where the resistance force on the sledge is only 2 N. She continues to pull the sledge with the same force as before and with the rope still at 30°.

(iii) Describe what happens to the sledge and to the girl.

9. A spaceship of mass 5000 kg is stationary in deep space. It fires its engines, producing a forward thrust of 2000 N for $2\frac{1}{2}$ minutes, and then turns them off.

(i) What is the speed of the spaceship at the end of the $2\frac{1}{2}$ minute period?

(ii) Describe the subsequent motion of the spaceship.

(iii) Illustrate the motion of the spaceship on a speed–time graph.

The spaceship then enters a cloud of interstellar dust which brings it to a halt after a distance of 7200 km.

(iv) What is the force of resistance (assumed constant) on the spaceship from the interstellar dust cloud?

The spaceship is travelling in convoy with another spaceship which is the same in all respects except that it is carrying an extra 500 kg of equipment. The second spaceship carries out exactly the same procedure as the first one.

(v) Which spaceship travels further into the dust cloud?

10. The picture shows a situation which has arisen between two anglers, Davies and Jones, standing at the ends of adjacent jetties. Their lines have become entangled under the water with the result that they have both hooked the same fish, which has mass 1.9 kg. Both are reeling in their lines as hard as they can in order to claim the fish.

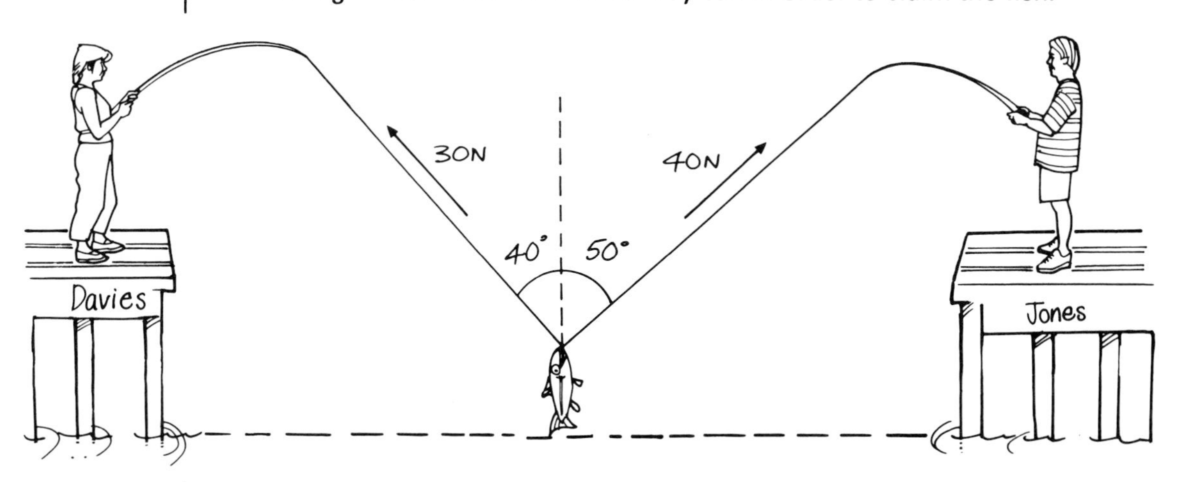

(i) Draw a diagram showing the forces acting on the fish.

(ii) Resolve the tensions in both anglers' lines into horizontal and vertical components and so find the total force acting on the fish.

(iii) Find the magnitude and direction of the acceleration of the fish.

At this point Davies' line breaks.

(iv) What happens to the fish?

Exercise 7D continued

11. Two cars collided head on. One had mass 1000 kg and was moving at 6 ms^{-1} and the other had mass 1500 kg. It took them 0.05 seconds to come to rest and they remained in contact (i.e. did not bounce off each other).

(i) Find the average acceleration experienced by the 1000 kg car.

(ii) Find the average force exerted on this car.

(iii) Find the average force on the heavier car.

(iv) Show that the heavier car must have been travelling at 4 ms^{-1}.

(v) What can you say about the damage to each car?

12. The diagram shows a builder, who works on his own, hauling a load of bricks up to the top of a building on a lift using a smooth pulley system. The weight of the builder is 800 N and that of the lift is 200 N.

Consider the lift at a position in mid-air.

(i) What is the greatest force the builder can exert on the rope?

(ii) What is the magnitude of the reaction of the ground on the builder when he is exerting this force?

(iii) What is the greatest weight of bricks that the builder can lift in this way?

One day the builder decides to ride on the lift himself as well.

(iv) What is the greatest weight of bricks he can now raise?

He enjoys riding on the lift so he attaches a pulley to the floor of the lift and passes the rope through it. He now pulls the rope upwards.

(v) What is the greatest weight of bricks he can now raise?

13. The position vector of a motorcycle of mass 150 kg on a track is modelled by

$$\mathbf{r} = 4t^2\mathbf{i} + \frac{t(8-t)^2}{8}\mathbf{j} \qquad 0 \leqslant t \leqslant 8$$

$$\mathbf{r} = (64t - 256)\mathbf{i} \qquad 8 < t \leqslant 20$$

where t is the time in seconds after the start of a race. The vectors **i** and **j** are in directions along and perpendicular to the direction of the track as shown in the diagram. The origin is in the middle of the track. The vector **k** has direction vertically upwards.

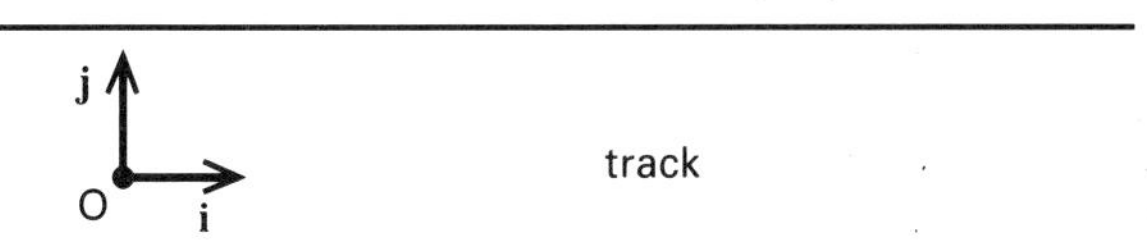

(i) Draw a sketch to show the motorcycle's path over the first 10 seconds. The track is 20 m wide. Does the motorcycle leave it?

(ii) Find, in vector form, expressions for the velocity and acceleration of the motorcycle at time t for $0 \leqslant t \leqslant 20$.

(iii) Find in vector form an expression for the resultant horizontal force acting on the motorcycle during the first 8 seconds, in terms of t.

(iv) Why would you expect the driving force from the motorcycle's engine to be substantially greater than the component in the **i** direction of your answer to part (iii)?

When $t = 22$ seconds the motorcycle's velocity is given by

$$\mathbf{v} = 60\mathbf{i} + 6\mathbf{k}.$$

(v) What has happened?

14. A spacecraft of mass 100 000 kg is stationary and without power in a part of space occupied by a constellation of 5 stars. The position vectors of these stars relative to the spacecraft, and their masses, are as follows:

Star	**Position Vector (light years)**	**Mass ($\times 10^{29}$ kg)**
α	$2\mathbf{i} + 2\mathbf{j} + \mathbf{k}$	54
β	$4\mathbf{i} + 3\mathbf{j}$	15
γ	$\mathbf{i}$	20
δ	$\mathbf{i} - 2\mathbf{j} - 2\mathbf{k}$	18
ε	$-3\mathbf{i} - 4\mathbf{k}$	20

The spacecraft is subject to a force of gravitational attraction, **F**, towards each of the 5 stars, where the magnitude of **F** is given in newtons by

Exercise 7D continued

$$F = \frac{KM_1M_2}{d^2}$$

where M_1 kg is the mass of the spacecraft, M_2 kg is the mass of the star in question, d is the distance between the star and the spacecraft and K is a constant. The effects of all other bodies may be ignored because they are much more distant.

(i) Find the distance of the spacecraft from the star α in light years.

(ii) Show that the magnitude of the force on the spacecraft from the star α is 6×10^{34} K newtons.

(iii) Show that the unit vector in the direction from the spacecraft to the star α is $\frac{2}{3}\mathbf{i} + \frac{2}{3}\mathbf{j} + \frac{1}{3}\mathbf{k}$ and hence that the force exerted on the spacecraft by the star α is $(4\mathbf{i} + 4\mathbf{j} + 2\mathbf{k}) \times 10^{34}$ K newtons.

(iv) Find, in vector form, the forces on the spacecraft from the other four stars.

(v) Hence find the resultant force acting on the spacecraft.

(vi) Given that in these units the value of the constant K is 7.45×10^{-43}, find the magnitude of the spacecraft's acceleration in ms^{-2}.

Investigations

An Accelerating Trolley

Set up the apparatus shown below.

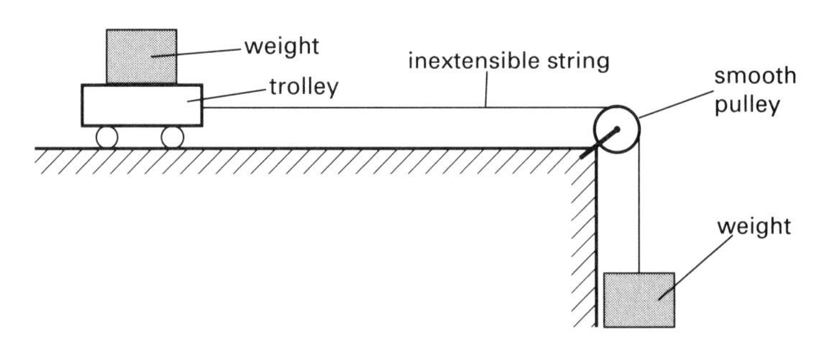

Measure the mass of the trolley (and any weights you load onto it) and that of the hanging weight. Calculate what you would expect the acceleration of the trolley to be when released from rest.

Time the trolley as it travels a measured distance across the table from rest, and so deduce its actual acceleration.

Compare your answer with your prediction. How well can you account for any discrepancy?

Resultant Forces

Set up the equipment shown below with the masses M_1, M_2 and M_3 all different.

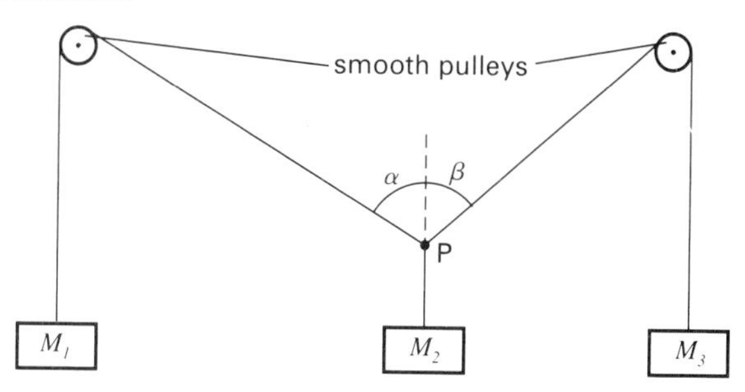

(i) Calculate the values of α and β you would expect at the equilibrium position for particular values of M_1, M_2 and M_3. Now let the system find its equilibrium position. Measure the angles α and β. Do your measurements agree with your predictions?

(ii) Calculate the direction in which you would expect the point P to move when it has been pulled away from its equilibrium position, for example you could choose a starting point to one side and downwards. Mount a felt pen so that it records the movement on a piece of paper behind the apparatus. Compare what happens with your predictions.

Historical Investigation

Investigate Aristotle's view of motion and explain how it differs from Newton's.

Describe three real situations which provide evidence to support Newton's view against Aristotle's.

What contribution did Galileo make to the study of motion?

KEY POINTS

- Newton's Laws of motion:

1st Law	Every body continues in a state of rest or uniform motion in a straight line unless acted on by an external force.
2nd Law	If a force **F** acts on a mass m, then it produces an acceleration **a** where $\mathbf{F} = m\mathbf{a}$.
3rd Law	To each force there is an equal and opposite reaction.
or	If two objects A and B exert forces on each other then the force exerted on A by B is equal in magnitude and opposite in direction to the force exerted on B by A.

- If the resultant of all the forces acting on a body is not zero the body will accelerate in the direction of the resultant.

8 Projectiles

Swift of foot was Hiawatha;
He could shoot an arrow from him,
And run forward with such fleetness,
That the arrow fell behind him!
Strong of arm was Hiawatha;
He could shoot ten arrows upwards,
Shoot them with such strength and swiftness,
That the last had left the bowstring,
Ere the first to earth had fallen!

The Song of Hiawatha, Longfellow

This stroboscopic photograph shows a ball moving as a projectile.

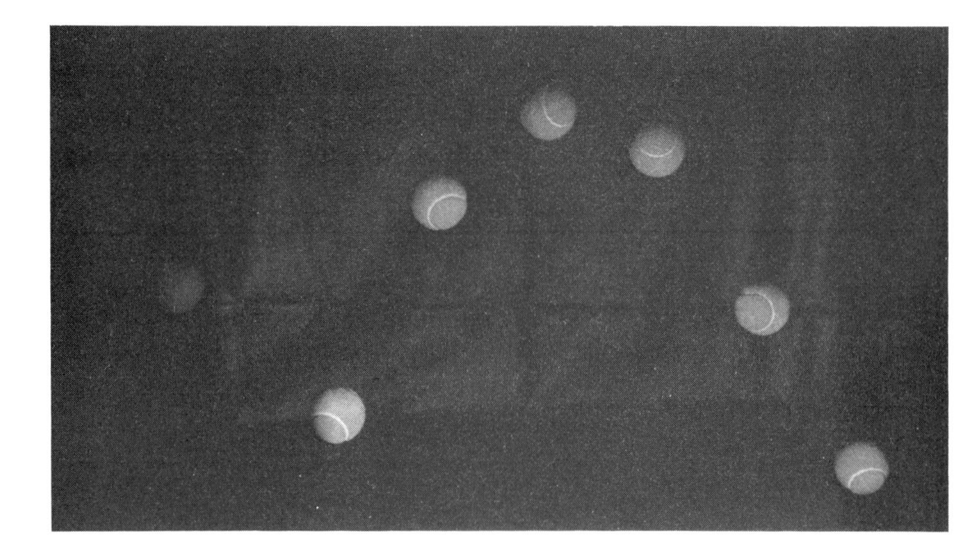

What are the main features of the motion?
Why is the ball called a projectile?

Experiment

When you throw a ball through the air it is difficult to see exactly what happens because it moves very quickly. This experiment allows you to simulate projectile motion happening more slowly.

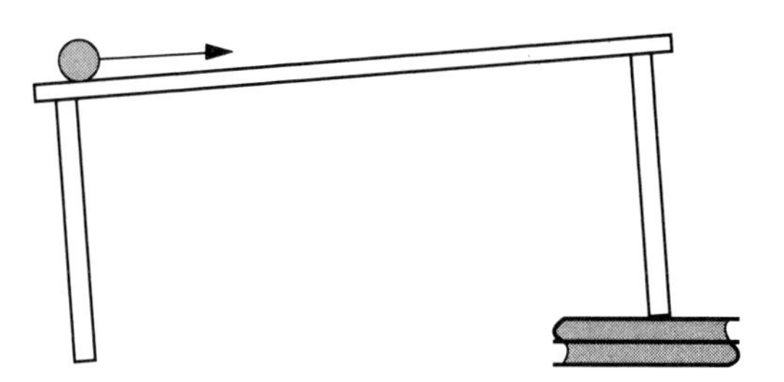

Tilt a table by putting books of equal thicknesses under two legs as shown in the diagram. Roll a ball up the table at different angles, using a length of tube or a v-shaped chute clamped in a retort stand as a launcher. Using such a launching system allows the ball to be launched at the same speed each time.

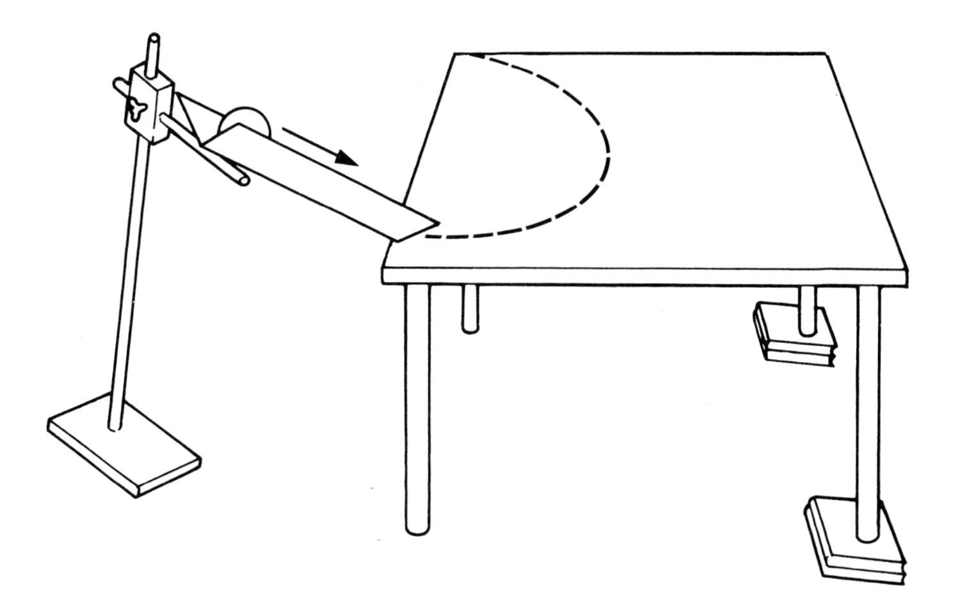

Before doing any experiments try to answer these questions intuitively.

- At what angle to the table edge should the ball be launched to make it travel the maximum distance (i.e. achieve its maximum *range*)?
- If the ball were released at a greater speed what would happen to its range?
- Is it possible to achieve the same range with two different angles?
- What would happen if the slope of the table were increased?
- What is the shape of the path that the ball describes?
- Does the range depend on the mass of the ball?

Now use the apparatus to investigate these questions.

Modelling projectile motion

The motion of a projectile can be modelled simply by making a number of important assumptions.

- The projectile is a particle.
- The projectile does not spin.
- The projectile stays close to the Earth's surface and so gravity is constant.
- There is no air resistance.

The following examples introduce you to the methods used.

EXAMPLE

A ball of mass 0.5 kg and diameter 0.05 m is thrown across horizontal ground, starting from ground level with speed 20 ms^{-1} at an angle of 60° to the horizontal. Find

(i) the position of the ball as a function of time,
(ii) the greatest height reached by the ball,
(iii) the time of flight,
(iv) the range of the ball.
(v) Sketch the path of the ball and mark on the sketch the information contained in your answers to parts (i) to (iv).

Commentary

Assuming that air resistance can be ignored, the only force acting on the ball during its flight is its own weight, and so the acceleration is g downwards. This has two consequences for subsequent work:

1 it is convenient to take components horizontally and vertically;
2 since the acceleration is constant, you may use the constant acceleration equations. (Notice, however, that if air resistance is considered, the acceleration is not constant and you must use calculus methods.)

In this example, the initial velocity, **u**, which is 20 ms^{-1} at 60° to the horizontal, can be written as:

Horizontal:
Vertical:
$$\begin{pmatrix} u_x \\ u_y \end{pmatrix} = \begin{pmatrix} 20\cos 60° \\ 20\sin 60° \end{pmatrix} = \begin{pmatrix} 10 \\ 17.3 \end{pmatrix} \text{in ms}^{-1}$$

The acceleration is

$$\mathbf{a} = \begin{pmatrix} 0 \\ -9.8 \end{pmatrix} \text{in ms}^{-2}$$

Initially the ball is at the origin.

Solution

(i) *The position at time t*

The displacement from the origin is given by

$$\mathbf{r} = \mathbf{u}t + \tfrac{1}{2}\mathbf{a}t^2$$

$$\begin{pmatrix} x \\ y \end{pmatrix} = \begin{pmatrix} 10 \\ 17.3 \end{pmatrix} t + \frac{1}{2}\begin{pmatrix} 0 \\ -9.8 \end{pmatrix} t^2$$

This can be written as two separate equations:

Horizontal: $x = 10t$

Vertical: $y = 17.3t - 4.9t^2$

(ii) *The maximum height*

When the ball is at maximum height, the vertical component of its velocity is zero.

The velocity of the ball is given by

$$\mathbf{v} = \mathbf{u} + \mathbf{a}t$$

Horizontally:
Vertically:
$$\begin{pmatrix} v_x \\ v_y \end{pmatrix} = \begin{pmatrix} 10 \\ 17.3 \end{pmatrix} + \begin{pmatrix} 0 \\ -9.8 \end{pmatrix} t \qquad \begin{matrix} : v_x = 10 \\ : v_y = 17.3 - 9.8t \end{matrix}$$

$$\text{When } v_y = 0, \ t = \frac{17.3}{9.8} = 1.77$$

Substituting this in the position expression gives

$$\begin{pmatrix} x \\ y \end{pmatrix} = \begin{pmatrix} 10 \times 1.77 \\ 17.3 \times 1.77 - 4.9 \times 1.77^2 \end{pmatrix} = \begin{pmatrix} 17.7 \\ 15.3 \end{pmatrix}$$

so the maximum height is 15.3 m.

(iii) *The time of flight*

The flight ends when the ball returns to the ground, that is when $y = 0$.

Substituting $y = 0$ in $\begin{pmatrix} x \\ y \end{pmatrix} = \begin{pmatrix} 10t \\ 17.3t - 4.9t^2 \end{pmatrix}$

gives

$$0 = 17.3t - 4.9t^2$$

$$t(4.9t - 17.3) = 0$$

$$t = 0 \text{ or } t = \frac{17.3}{4.9} = 3.53$$

Now $t = 0$ is the time when the ball is thrown;

$t = 3.53$ is the time when the ball lands.

The flight time is 3.53 seconds.

(iv) *The range of the ball*

The range of the ball is the horizontal distance it travels before landing. This is found by substituting $t = 3.53$ in

$$\begin{pmatrix} x \\ y \end{pmatrix} = \begin{pmatrix} 10t \\ 17.3t - 4.9t^2 \end{pmatrix}$$

This gives

$$\begin{pmatrix} x \\ y \end{pmatrix} = \begin{pmatrix} 10 \times 3.53 \\ 17.3 \times 3.53 - 4.9 \times 3.53 \end{pmatrix} = \begin{pmatrix} 35.3 \\ 0 \end{pmatrix}$$

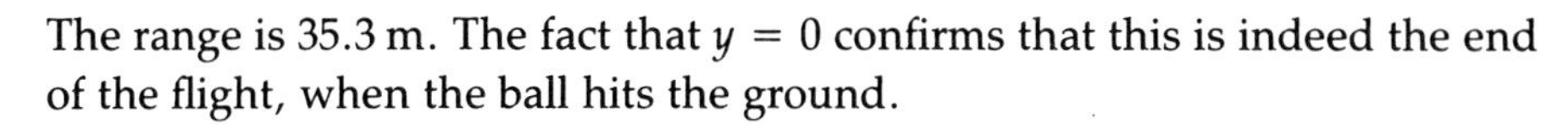

The range is 35.3 m. The fact that $y = 0$ confirms that this is indeed the end of the flight, when the ball hits the ground.

(v) *The trajectory*

At a general time t,

$$\begin{pmatrix} x \\ y \end{pmatrix} = \begin{pmatrix} 10t \\ 17.3t - 4.9t^2 \end{pmatrix}$$

gives the position.

Highest point of flight is 15.3 m when $t = 1.77$. This is half way through the flight.

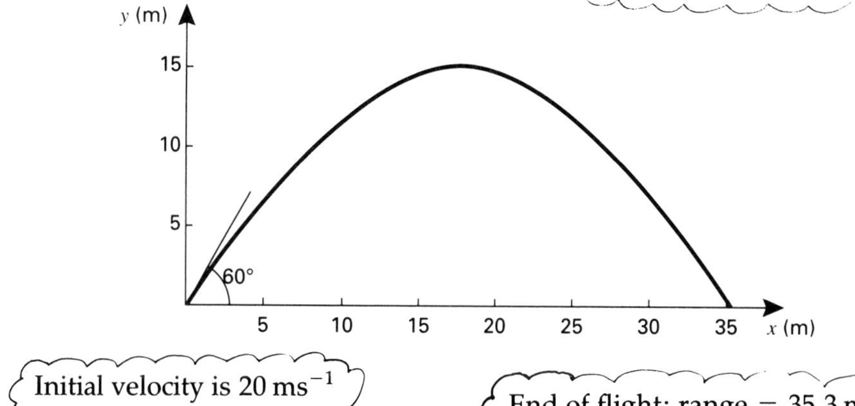

Initial velocity is 20 ms^{-1} at 60° to the horizontal,

i.e. $\begin{pmatrix} 10 \\ 17.3 \end{pmatrix}$ ms^{-1}.

End of flight: range = 35.3 m, time = 3.53 s.

Notice that the curve is a parabola, symmetrical about the vertical line through its highest point.

EXAMPLE

A girl throws a stone horizontally out to sea at 20 ms^{-1} from the top of a cliff 40 m high. How far from the base of the cliff does the stone hit the sea?

Solution

It is essential to define the origin for this question carefully. It could be the point of projection or the base of the cliff, but you must make a decision and keep to it. In this solution it is taken to be the base of the cliff.

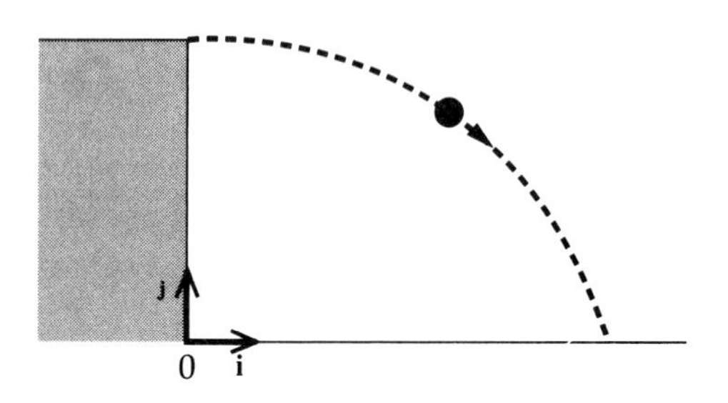

Initial position $= \begin{pmatrix} 0 \\ 40 \end{pmatrix}$; initial velocity $= \begin{pmatrix} 20 \\ 0 \end{pmatrix}$;

acceleration $= \begin{pmatrix} 0 \\ -9.8 \end{pmatrix}$

Using $\mathbf{r} = \mathbf{r}_0 + \mathbf{u}t + \frac{1}{2}\mathbf{a}t^2$

$$\begin{pmatrix} x \\ y \end{pmatrix} = \begin{pmatrix} 0 \\ 40 \end{pmatrix} + \begin{pmatrix} 20 \\ 0 \end{pmatrix} t + \frac{1}{2}\begin{pmatrix} 0 \\ -9.8 \end{pmatrix} t^2$$

When the stone hits the sea, $y = 0$

$$0 = 40 - 4.9t^2$$

$$t = \sqrt{\frac{40}{4.9}} = 2.86$$

When $t = 2.86$, $x = 20 \times 2.86 = 57$ (to the nearest whole number).
The stone hits the sea 57 m from the base of the cliff.

For Discussion

Describe three real situations that could be described as 'projectile motion'. What assumptions should you make in each case? How valid are your assumptions?

Exercise 8A

1. A ball is thrown from a point at ground level with velocity 20 ms^{-1} at 30° to the horizontal. The ground is level and horizontal and you should ignore air resistance. Take g to be 10 ms^{-2}.

(i) Find the horizontal and vertical components of the ball's initial velocity.

(ii) Find the horizontal and vertical components of the ball's acceleration.

(iii) Find the horizontal distance travelled by the ball before its first bounce.

(iv) Find how long the ball takes to reach maximum height.

(v) Find the maximum height reached by the ball.

2. Nick hits a golf ball with initial velocity 50 ms^{-1} at 35° to the horizontal.

(i) Find the horizontal and vertical components of the ball's initial velocity.

(ii) Calculate the position of the ball at one second intervals for the first six seconds of its flight.

(iii) Draw a graph of the ball's trajectory and use it to estimate

(a) the maximum height of the ball,

(b) the horizontal distance the ball travels before bouncing.

Exercise 8A continued

(iv) Calculate the maximum height the ball reaches and the horizontal distance it travels before bouncing, and compare your answers with the estimates you found from your graph.

(v) State the modelling assumptions you made in answering this question.

3. Claire scoops a hockey ball off the ground, giving it an initial velocity of 19 $\mathrm{ms^{-1}}$ at 25° to the horizontal.

(i) Find the horizontal and vertical components of the ball's initial velocity.

(ii) Find the time that elapses before the ball hits the ground.

(iii) Find the distance the ball travels before hitting the ground.

(iv) Find how long it takes for the ball to reach maximum height.

(v) Find the maximum height reached.

A member of the opposing team is standing 20 m away from Claire.

(vi) How high is the ball when it passes her? Can she stop the ball?

4. A footballer is standing 30 m in front of the goal. He kicks the ball towards the goal with velocity 18 $\mathrm{ms^{-1}}$ and angle 55° to the horizontal. The height of the goal's crossbar is 2.5 m. Air resistance and spin may be ignored.

(i) Find the horizontal and vertical components of the ball's initial velocity.

(ii) Find the time it takes for the ball to cross the goal-line.

(iii) Does the ball bounce in front of the goal, go straight into the goal or go over the crossbar?

In fact the goalkeeper is standing 5 m in front of the goal and will stop the ball if its height is less than 2.8 m when it reaches him.

(iv) Does the goalkeeper stop the ball?

5. An airliner is flying at a speed of 300 $\mathrm{ms^{-1}}$ and maintaining an altitude of 10 000 m when a bolt becomes detached from it. Ignoring air resistance, find

(i) the time that the bolt takes to reach the ground,

(ii) the horizontal distance between the point where the bolt leaves the airliner and the point where it hits the ground,

(iii) the speed of the bolt when it hits the ground,

(iv) the angle to the horizontal at which the bolt hits the ground.

Exercise 8A continued

6. Reena is learning to serve in tennis. She hits the ball from a height of 2 metres. For her serve to be legal it must pass over the net which is 12 m away from her and 0.91 m high, and it must land within 6.4 m of the net.

Make the following modelling assumptions to answer the questions.
(a) She hits the ball horizontally.
(b) Air resistance may be ignored.
(c) The ball may be treated as a particle.
(d) The ball does not spin.
(e) She hits the ball straight down the middle of the court.

(i) How long does the ball take to fall to the level of the top of the net?
(ii) How long does the ball take from being hit to first reaching the ground?
(iii) What is the lowest speed with which Reena must hit the ball to clear the net?
(iv) What is the greatest speed with which she may hit it if it is to land within 6.4 m of the net?
(v) Explain how you know that these answers are unrealistic, and state which of the modelling assumptions you would wish to change.

7. Sharon is diving into a swimming pool. During her flight she may be modelled as a particle, with initial velocity 1.8 ms^{-1} at angle 30° above the horizontal, and initial position 3.1 m above the water. Air resistance may be neglected.
(i) Find the greatest height above the water that Sharon reaches during her dive.
(ii) Show that the time t, in seconds, that it takes Sharon to reach the water is given by

$$4.9t^2 - 0.9t - 3.1 = 0$$

and solve this equation to find t. Explain the significance of the other solution to the equation.

Just as Sharon is diving a small boy jumps into the swimming pool, hitting the water at a point in line with the diving board and 1.5 m from its end.
(iii) Is there an accident?

8. A stunt motorcycle rider attempts to jump over a gorge 50 m wide. He uses a ramp at 25° to the horizontal for his take-off and has a speed of 30 ms^{-1} at this time.
(i) Assuming that air resistance is negligible, find out whether the rider crosses the gorge successfully.

The stunt man actually believes that in any jump the effect of air resistance is to reduce his distance by 40%.
(ii) Calculate his minimum safe take-off speed for this jump.

Exercise 8A continued

9. To kick a goal in rugby you must kick the ball over the crossbar of the goal posts (height 3.0 m), between the two uprights. Dafyd Evans attempts a kick from a distance of 35 m. The initial velocity of the ball is 20 ms^{-1} at 30° to the horizontal. The ball is aimed between the uprights and no spin is applied.

(i) How long does it take the ball to reach the goal posts?

(ii) Does it go over the crossbar?

Later in the game, Dafyd takes another kick from the same position and hits the crossbar.

(iii) Given that the initial velocity of the ball in this kick was also at 30° to the horizontal, find the initial speed.

Many rugby kickers choose to give the ball spin.

(iv) What effect does spin have upon the flight of the ball? Discuss briefly the advantages and disadvantages of using spin.

10. In this question take g to be 10 ms^{-2}. A catapult projects a small pellet at speed 20 ms^{-1} and can be directed at any angle to the horizontal.

(i) Find the range of the catapult when the angle of projection is
(a) 30° (b) 40° (c) 45° (d) 50° (e) 60°

(ii) Show algebraically that the range is the same when the angle of projection is α as it is when the angle is $90° - \alpha$.

The catapult is angled with the intention that the pellet should hit a point on the ground 36 m away.

(iii) Show that one appropriate angle of projection would be 32.1° and write down another suitable angle.

In fact the angle of projection from the catapult is liable to error.

(iv) Find the distance by which the pellet misses the target in each of the cases in (iii) when the angle of projection is subject to an error of +0.5°. Which angle should you use for greater accuracy?

11. A cricketer hits the ball on the half-volley, i.e. when the ball is at ground level. The ball leaves the ground at an angle of 30° to the horizontal and travels towards a fielder standing on the boundary 60 m away.

(i) Find the initial speed of the ball if it hits the ground for the first time at the fielder's feet.

(ii) Find the initial speed of the ball if it is at a height of 3.2 m (well outside the fielder's reach) when it passes over the fielder's head.

In fact the fielder is able to catch the ball without moving provided that its height, h metres, when it reaches him satisfies the inequality $0.25 \leqslant h \leqslant 2.1$.

(iii) Find a corresponding range of values for u, the initial speed of the ball.

12. A horizontal tunnel has a maximum height of 3 m. A ball is thrown inside the tunnel with an initial speed of 18 ms^{-1}. What is the greatest horizontal distance that the ball can travel before it bounces for the first time?

The path of a projectile

It can be useful to describe the path of a projectile by giving its equation in terms of x and y.

EXAMPLE

A projectile is launched with an initial velocity 20 ms^{-1} at an angle of 30° to the horizontal.

(i) Write down the position vector of the projectile, if it is launched at the origin.

(ii) Show that the equation of the path is the parabola

$$y = 0.578x - 0.016x^2.$$

Solution

(i) $\mathbf{u} = \begin{pmatrix} 20\cos 30° \\ 20\sin 30° \end{pmatrix}, \quad \mathbf{a} = \begin{pmatrix} 0 \\ -9.8 \end{pmatrix}, \quad \mathbf{r}_0 = \begin{pmatrix} 0 \\ 0 \end{pmatrix}$

Using $\quad \mathbf{r} = \mathbf{u}t + \frac{1}{2}\mathbf{a}t^2$

$\Rightarrow \quad \mathbf{r} = \begin{pmatrix} x \\ y \end{pmatrix} = \begin{pmatrix} 20\cos 30° \\ 20\sin 30° \end{pmatrix} t + \frac{1}{2}\begin{pmatrix} 0 \\ -9.8 \end{pmatrix} t^2$

$\therefore \quad \mathbf{r} = \begin{pmatrix} x \\ y \end{pmatrix} = \begin{pmatrix} 17.3t \\ 10t - 4.9t^2 \end{pmatrix}$

(ii) To obtain the equation of the path we need to eliminate t between the two equations

$$x = 17.3t \quad \text{and} \quad y = 10t - 4.9t^2.$$

From the first of these equations,

$$t = \frac{x}{17.3}.$$

Substituting this into the equation for y gives

$$y = 10\left(\frac{x}{17.3}\right) - 4.9\left(\frac{x}{17.3}\right)^2$$

$$\Rightarrow \quad y = 0.578 - 0.016x^2.$$

In the next example, the work done earlier in this chapter is repeated for the general case. A number of standard results and formulae are obtained.

EXAMPLE

A particle is projected from the origin with speed u at an angle α to the horizontal. Throughout its motion the only force acting on the particle is the force due to gravity. The x and y axes are horizontal and vertical through the origin, O, in the plane of motion of the particle.

(i) Show that the position vector is modelled by $\mathbf{r} = x\mathbf{i} + y\mathbf{j}$, where

$$x = ut\cos\alpha \quad \text{and} \quad y = ut\sin\alpha - \tfrac{1}{2}gt^2.$$

(ii) Eliminate t to show that the projectile follows the parabolic path

$$y = x\tan\alpha - \frac{gx^2}{2u^2}\sec^2\alpha.$$

(iii) Find the time taken for the particle to reach its greatest height and show that this height is

$$\frac{u^2\sin^2\alpha}{2g}.$$

(iv) Show that the time of flight of the particle is

$$\frac{2u^2\sin\alpha}{g}.$$

(v) Show that the horizontal range of the particle is

$$\frac{2u^2\sin\alpha\cos\alpha}{g}.$$

What is the maximum horizontal range for different values of α?

Solution

(i) Acceleration, $\mathbf{a} = -g\mathbf{j}$; initial velocity $\mathbf{u} = u\cos\alpha\mathbf{i} + u\sin\alpha\mathbf{j}$.
The acceleration is constant, so constant acceleration equations may be used.

Using $\quad \mathbf{r} = \mathbf{u}t + \tfrac{1}{2}\mathbf{a}t^2 \quad$ gives

$$\mathbf{r} = ut\cos\alpha\mathbf{i} + (ut\sin\alpha - \tfrac{1}{2}gt^2)\mathbf{j}$$

Since $\mathbf{r} = x\mathbf{i} + y\mathbf{j}$,

$x = ut\cos\alpha$ and $y = ut\sin\alpha - \tfrac{1}{2}gt^2$, as required.

(ii) From the first of these equations

$$t = \frac{x}{u\cos\alpha}.$$

This can be substituted into $y = ut\sin\alpha - \frac{1}{2}gt^2$ to give

$$y = u\left(\frac{x}{u\cos\alpha}\right)\sin\alpha - \tfrac{1}{2}g\left(\frac{x}{u\cos\alpha}\right)^2$$

$$= x\frac{\sin\alpha}{\cos\alpha} - \frac{gx^2}{2u^2\cos^2\alpha}$$

$$= x\tan\alpha - \frac{gx^2}{2u^2}\sec^2\alpha$$

$\sec\alpha = \dfrac{1}{\cos\alpha}$

(iii) The velocity is $\mathbf{v} = \mathbf{u} + \mathbf{a}t$

$$= u\cos\alpha\mathbf{i} + (u\sin\alpha - gt)\mathbf{j}$$

When the projectile reaches its greatest height, the vertical component of its velocity is zero, so

$$u\sin\alpha - gt = 0$$

giving $\quad t = \dfrac{u\sin\alpha}{g}.$

The height of the projectile is given by

$$y = ut\sin\alpha - \tfrac{1}{2}gt^2,$$

so the greatest height is

$$y = u\left(\frac{u\sin\alpha}{g}\right)\sin\alpha - \tfrac{1}{2}g\left(\frac{u\sin\alpha}{g}\right)^2$$

$$= \frac{u^2\sin^2\alpha}{g} - \frac{u^2\sin^2\alpha}{2g}$$

$$= \frac{u^2\sin^2\alpha}{2g}.$$

(iv) When the projectile hits the ground, $y = 0$. So

$$0 = ut\sin\alpha - \tfrac{1}{2}gt^2$$

$$= t(u\sin\alpha - \tfrac{1}{2}gt).$$

This equation has two solutions, $t = 0$ and $t = \dfrac{2u\sin\alpha}{g}$.

The first corresponds to the start of the motion and the second to when the projectile hits the ground.

(v) The range of the projectile is the value of x when

$$t = \frac{2u\sin\alpha}{g}.$$

$$\text{Now } x = ut\cos\alpha, \text{ so range} = u\left(\frac{2u\sin\alpha}{g}\right)\cos\alpha$$

$$= \frac{2u^2\sin\alpha\cos\alpha}{g}.$$

Note that $2\sin\alpha\cos\alpha = \sin(2\alpha)$, so the range can be expressed as

$$\text{range} = \frac{u^2\sin(2\alpha)}{g}$$

The range is a maximum when $\sin(2\alpha) = 1$, that is when $2\alpha = 90°$ or $\alpha = 45°$.

Exercise 8B

In all questions in this exercise, unless otherwise stated, make the simplification that g is 10 ms^{-2} and use the modelling assumptions that air resistance can be ignored and the ground is horizontal.

1. A particle is projected with initial velocity 50 ms^{-1} at an angle of 36.9° to the horizontal. The point of projection is taken to be the origin, with the x axis horizontal and the y axis vertical in the plane of the particle's motion.

(i) Show that at time t seconds, the height of the particle in metres is given by

$$y = 30t - 5t^2$$

and write down the corresponding expression for x.

(ii) Eliminate t between your equations for x and y to show that

$$y = \frac{3x}{4} - \frac{x^2}{320}.$$

(iii) Plot the graph of $y = \frac{3x}{4} - \frac{x^2}{320}$ using a scale of 2 cm for 10 m along both axes.

(iv) Mark in the points corresponding to the position of the particle after 1, 2, 3, 4 . . . seconds.

2. A golfer hits the ball with initial velocity 50 ms^{-1} at an angle α to the horizontal where $\sin\alpha = 0.6$.

(i) Find the equation of its trajectory, assuming that air resistance may be neglected.

The flight of the ball is recorded on film and its position vector, from the point where it was hit, is calculated. The unit vectors **i** and **j** are horizontal and vertical in the plane of the ball's motion. The results (to the nearest 0.5 m) are as follows:

Time (s)	Position (m)	Time (s)	Position (m)
0	$0\mathbf{i} + 0\mathbf{j}$	4	$152\mathbf{i} + 39\mathbf{j}$
1	$39.5\mathbf{i} + 24.5\mathbf{j}$	5	$187.5\mathbf{i} + 24.5\mathbf{j}$
2	$78\mathbf{i} + 39\mathbf{j}$	6	$222\mathbf{i} + 0\mathbf{j}$
3	$116.5\mathbf{i} + 44\mathbf{j}$		

(ii) On the same piece of graph paper draw the trajectory you found in (i) and that found from analysing the film. Compare the two graphs and suggest a reason for any differences.

(iii) It is suggested that the horizontal component of the resistance to the motion of the golf ball is almost constant. Are the figures consistent with this? Why is it reasonable to think the horizontal component of this force might be nearly constant but not the vertical component?

3. Melissa is playing volleyball. She hits the ball with initial speed u ms^{-1} from a height of 1 m at an angle 30° to the horizontal.

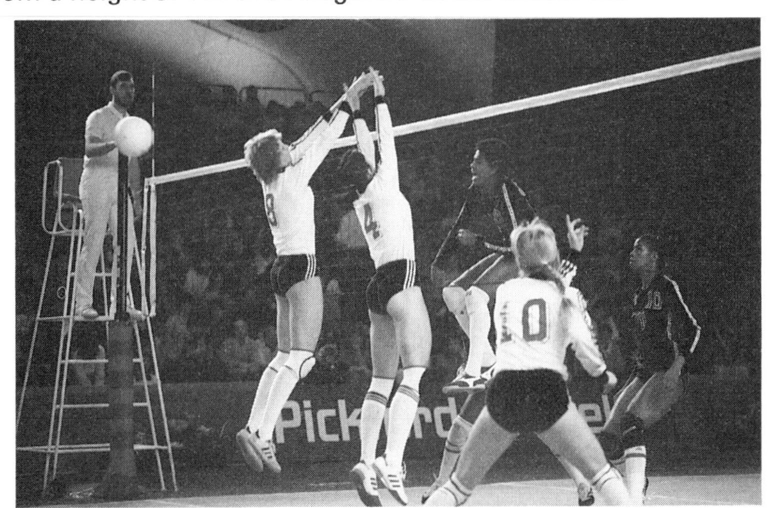

(i) Define a suitable origin and x and y axes, and find the equation of the trajectory of the ball in terms of x, y and u.

The rules of the game require the ball to pass over the net which is at height 2 m and land inside the court on the other side, which is of length 5 m. Melissa hits the ball straight along the court and is 3 m from the net when she does so.

(ii) Find the minimum value of u for the ball to pass over the net.

(iii) Find the maximum value of u for the ball to land inside the court.

(iv) What advice would you give Melissa?

4. While practising his tennis serve, Matthew hits the ball from a height of 2.5 m with a velocity of magnitude 25 ms^{-1} at an angle of 5° above the horizontal as shown in the diagram.
In this question, take $g = 9.8\,\text{ms}^{-2}$.

Exercise 8B continued

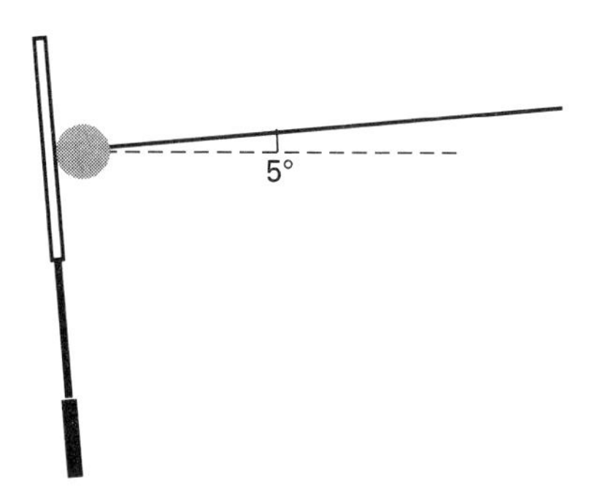

(i) Show that while in flight

$$y = 2.5 + 0.087x - 0.0079x^2$$

(ii) Find the horizontal distance from the serving point to the spot where the ball lands.

(iii) Determine whether the ball would clear the net, which is 1 m high and 12 m from the serving position in the horizontal direction.

5. A particle is projected up a slope of angle β where $\tan\beta = \frac{1}{2}$. The initial velocity of the particle is 50 ms^{-1} at angle α to the horizontal where $\sin\alpha = 0.8$. The x and y axes are taken from the point of projection and are in the plane of the particle's motion as shown below.

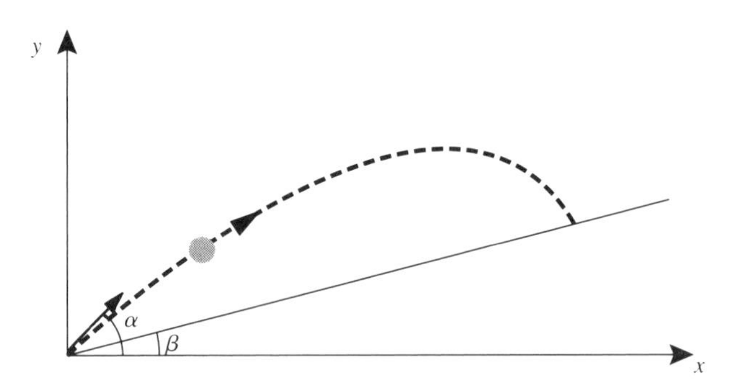

(i) Find the equation of the trajectory of the particle in terms of x, y, g and α.

(ii) Explain why the equation of the slope is $y = \frac{1}{2}x$.

(iii) Solve the equations of the trajectory and the slope to find the coordinates of the point where the particle hits the slope.

(iv) What is the range of the particle up the slope?

Another particle is then projected from the same point and with the same initial speed but at an angle of 45° to the horizontal.

(v) Find the range of this particle up the slope.

(vi) Does an angle of projection of 45° to the horizontal result in the particle travelling the maximum distance up the slope?

6. A golf ball is driven from the tee with speed $30\sqrt{2}\,\text{ms}^{-1}$ at an angle α to the horizontal.
   (i) Show that during its flight the horizontal and vertical displacements x and y of the ball from the tee satisfy the equation

$$y = x\tan\alpha - \frac{x^2}{360}\sec^2\alpha.$$

   (Note: $\sec\alpha = 1/\cos\alpha$; $\sec^2\alpha = 1 + \tan^2\alpha$.)
   (ii) The golf ball just clears a tree 5 m high which is 150 m horizontally from the tee. Find the two possible values of $\tan\alpha$.
   (iii) Find the greatest distance by which the golf ball can clear the tree for different angles of projection α, and find the value of $\tan\alpha$ in this case.
   (iv) The ball is aimed at the hole which is on the green immediately behind the tree. The hole is 160 m from the tee. What is the greatest height the tree could be without making it impossible to hit a 'hole in one'?

Investigations

Serving a tennis ball

When a tennis ball is served it must pass over the net and land inside the service court. Construct a mathematical model of the flight of a successful serve.

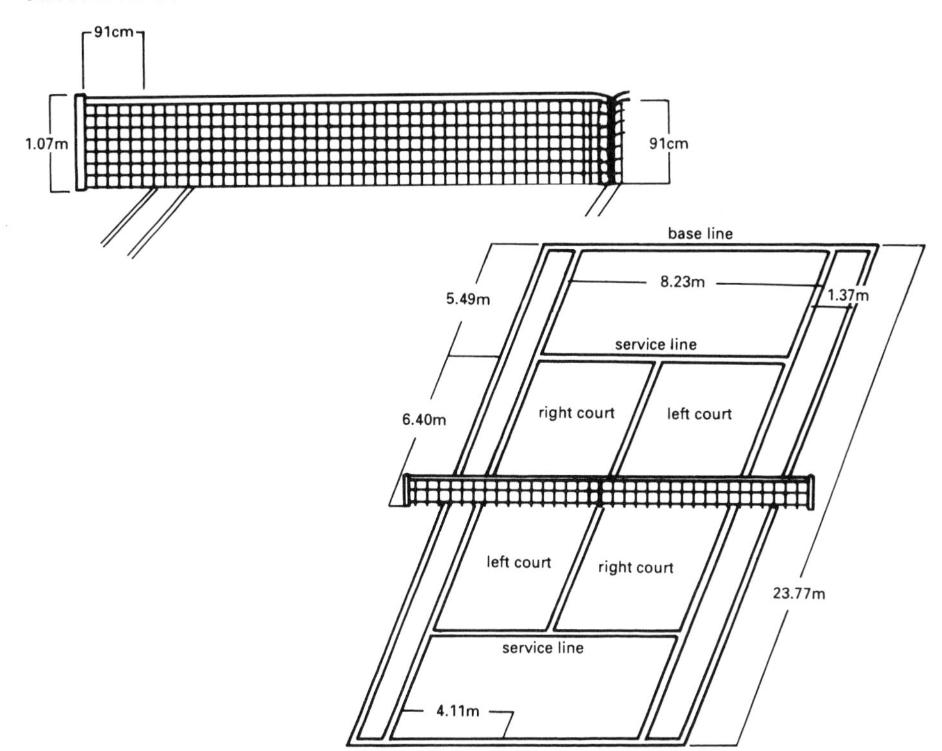

Shot putter

At what angle should a shot putter project the shot for best results?

Volcanoes

Volcanoes eject rocks at speeds of up to 700 ms^{-1}, in all directions. A scientist wishes to film an erupting volcano from as close as she can without endangering the lives of the crew of the light aircraft she is using. Define a 'danger zone' into which the aircraft should not fly.

Low level bombing

One of the dangers faced by the crew of a low level bomber is that they get caught in the blast from their own bomb. Construct a mathematical model to show both how this could happen and how it could be avoided.

Trajectories

The modelling in this chapter has led you to the conclusion that the trajectory of a projectile is a parabola. Devise and carry out experiments to investigate whether this really is the case.

Juggler

A juggler works with a number of balls. He throws each one up in the air with his right hand, catches it in his left hand and throws it back to his right hand. Construct a mathematical model for this, starting with the case of 1 ball, then going on to 2, 3, . . . etc.

KEY POINTS

- Projectile motion is usually considered in terms of horizontal and vertical components.

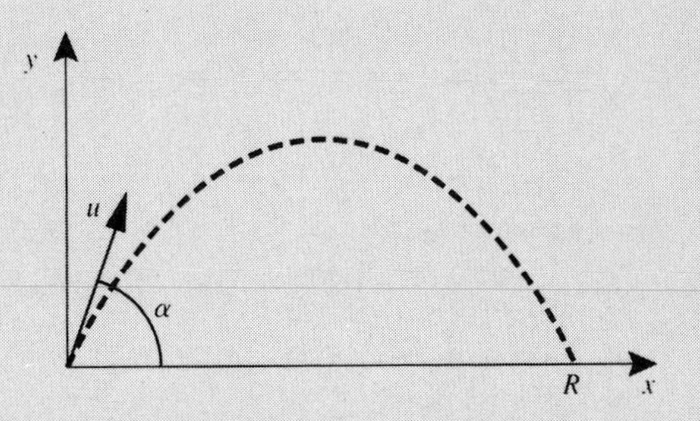

Angle of projection = α
Range = R

$$\text{Initial velocity } \mathbf{u} = \begin{pmatrix} u\cos\alpha \\ u\sin\alpha \end{pmatrix}$$

$$\text{Acceleration } \mathbf{g} = \begin{pmatrix} 0 \\ -g \end{pmatrix}$$

- At time t, $\mathbf{v} = \mathbf{u} + \mathbf{a}t$: $\begin{pmatrix} v_x \\ v_y \end{pmatrix} = \begin{pmatrix} u\cos\alpha \\ u\sin\alpha \end{pmatrix} + \begin{pmatrix} 0 \\ -g \end{pmatrix} t$

  and $\mathbf{s} = \mathbf{s}_0 + \mathbf{u}t + \frac{1}{2}\mathbf{a}t^2$; $\begin{pmatrix} x \\ y \end{pmatrix} = \begin{pmatrix} x_0 \\ y_0 \end{pmatrix} + \begin{pmatrix} u\cos\alpha \\ u\sin\alpha \end{pmatrix} t + \frac{1}{2}\begin{pmatrix} 0 \\ -g \end{pmatrix} t^2$

- At maximum height $v_y = 0$.
- On horizontal ground $y = 0$ when the projectile lands.
- If the projectile starts and ends on horizontal ground the range is maximum for an angle of projection of 45°.

Answers to selected exercises

2 KINEMATICS

Exercise 2A

1. $-4, 0, 5$; (i) 4 (ii) -5
2. (i) 2, 0, 0, 2 (iii) +10 (iv) $14\frac{1}{2}$
4. (iii) (A) 2 (B) 7
5. (i) $-16, -20, 0, 56$ (iii) 0, negative; 3, positive
6. (i) $t = 1$ (ii) -4 (iii) 4.08

Exercise 2B

1. (i) 59.887 (ii) 59.425
2. (i) 20 mph (ii) 23.7 mph
3. 774.2 mph
4. 1.753 ms^{-2}
5. 3.09 ms^{-1}

Exercise 2C

7. (i) Never (ii) Never
8. (i) 1 (ii) $10t$ (iii) 10ms^{-1}
9. (i) 125m (ii) 34.5 mm^{-1} (iii) 7.42 s
10. 2.5×10^3 cms^{-1}

Exercise 2D

1. 85 m
2. (i) J catches S (ii) T orders, S pays
3. (ii) $20 + \frac{1}{12}t$ (iii) 360 s (iv) 3940 m
4. 3562.5 m
5. 558 m
6. (i) When $t = 6$ (ii) 972 m
7. (i) 4.47 s (ii) 119 m
8. (i) $15 - 10t$ (ii) 11.5 m, + 5 ms^{-1}, 5 ms^{-1}; 11.5 m, – 5 ms^{-1}, 5 ms^{-1} (iv) When $t = 3$
9. (i) –3 m, –1 ms^{-1}, 1 ms^{-1} (ii) (a) 1s (b) 2.15 s
10. (i) $16t - 4t^2$ (ii) (a) 2 s (b) 4 s
11. (i) When $t = 1.13$ (ii) $2 - 9.8t$ (iii) 9.07 ms^{-1}
12. When $t = 2$

Exercise 2E

1. (i) 10, –10 (ii) 0, 36
2. 1 ms^{-2}
3. $s = 4t + 2t^2 - \frac{1}{3}t^3$ $v = 4 + 4t - t^2$
4. (i) D (ii) B, C, E (iii) A
6. (i) $v = 10t + \frac{3}{2}t^2 - \frac{1}{3}t^3$, $x = 5t^2 + \frac{1}{2}t^3 - \frac{1}{12}t^4$ (ii) $v = 2 + 2t^2 - \frac{2}{3}t^3$, $x = 1 + 2t + \frac{2}{3}t^3 - \frac{1}{6}t^4$ (iii) $v = -5 + 10t - t^2$, $x = -5t + 5t^2 - \frac{1}{3}t^3$
7. (iii) $-3t + 12$ (ms^{-1}) (iv) 0 ms^{-1} (v) –3 ms^{-2}
8. (i) A 10 ms^{-1}, E 9.6 ms^{-1} (iii) 11.52 m (iv) 11.62 s (v) E, 0.05 s, 0.47 m (vi) A
9. (ii) 1092 m (iii) 8.5 ms^{-2}, $1.6(t - 20)$, 0, 16 ms^{-2}
10. (ii) –0.000025, 0.05 (iii) 50 ms^{-1} (iv) 0 (v) $111\frac{2}{3}$ km

Exercise 2F

1. 5 ms^{-1}, 30 m
2. $\frac{25}{12}$ ms^{-2}, 150 m

3. (i) 360 m, 30 s, (ii) 250 m, 3 ms^{-2} (iii) 32 m, 16 ms^{-1} (iv) 16 ms^{-1}, 2 s (v) 6 ms^{-1} 1 ms^{-2}
4. $4\frac{1}{2}$ ms^{-2}
5. –8 ms^{-2}, 3 s
6. 604.9 s, 9036.6 m
7. (i) $2 + 0.4t$ (ii) $2t + 0.2t^2$ (iii) 18 ms^{-1}
8. No
9. (ii) $15t - 5t^2$ (iii) $30 - 5t^2$ (iv) When $t = 2$ (v) 10 m
10. (ii) 4 ms^{-1} (iii) At 2 s, 6 s; 0.1 ms^{-2} (iv) Approx, 2.7 ms^{-2}
11. (i) 5.4 ms^{-1} (ii) 4.4 ms^{-1} (iii) 1 ms^{-1} increase (iv) 9 ms^{-1}
12. 7.5 m
13. $43\frac{3}{4}$ m

3 VECTORS

Exercise 3A

2. **f, h, d**
3. 3 m, +50°; 3 m, –40°; 1 m, 0°
4. 3.7 km, 054°
5. 12.8 km, 039°
6. 10 m, 53° to horizontal
7. 080°
9. (i) $\overrightarrow{AB} = 2\overrightarrow{DC}$ (ii) $\overrightarrow{AB} = -2\overrightarrow{CD}$ (ii) $\overrightarrow{EA} = \frac{7}{10}\overrightarrow{CB}$ (iv) $\overrightarrow{AE} = -\frac{7}{10}\overrightarrow{CB}$
11. (i) $-\mathbf{p} + \mathbf{q}$ (ii) $-\frac{1}{3}\mathbf{p} + \frac{1}{3}\mathbf{q}$ (iii) $\frac{2}{3}\mathbf{p} + \frac{1}{3}\mathbf{q}$ (iv) $\frac{1}{3}\mathbf{p} + \frac{1}{6}\mathbf{q}$
12. (i) $\mathbf{a} + \mathbf{c}$ (ii) $-\mathbf{a} + \mathbf{c}$ (iii) $\mathbf{a} - \mathbf{c}$ (iv) $-\mathbf{a} - \mathbf{c}$ (v) $\frac{1}{2}\mathbf{c}$ (vi) $\mathbf{a} + \frac{1}{2}\mathbf{c}$ (vii) $-\mathbf{a} + \frac{1}{2}\mathbf{c}$
13. (i) $-\mathbf{p} + \mathbf{q}$ (ii) $-\mathbf{p}$ (iii) $-\mathbf{p} + \mathbf{q}$ (iv) $-2\mathbf{p} + \mathbf{q}$ (v) $-3\mathbf{p} + 2\mathbf{q}$ (vi) $-2\mathbf{p} + \mathbf{q}$
14. 045° until SE of B, then 315°

Exercise 3B

1. (i) E 6 m, N 2 m (ii) E 6 m, N 2 m (iii) E 6 m N 4 m
2. $\mathbf{a} = -2\mathbf{i}$; $\mathbf{b} = \mathbf{j}$; $\mathbf{c} = -3\mathbf{i}$; $\mathbf{d} = 3\mathbf{j}$; $\mathbf{e} = 2\mathbf{i}$; $\mathbf{f} = \mathbf{i} + \mathbf{j}$; $\mathbf{g} = -2\mathbf{i} - \mathbf{j}$; $\mathbf{h} = \mathbf{i} - 2\mathbf{j}$; $\mathbf{k} = \mathbf{i} - \mathbf{j}$
3. (4, – 11)
4. (i) $9\mathbf{i} - 3\mathbf{j}$ (ii) $4\mathbf{i} - 2\mathbf{j}$ (iii) $-7\mathbf{i} + 2\mathbf{j}$ (iv) $4\mathbf{i} - 6\mathbf{j}$ (v) $-3\mathbf{i} + 11\mathbf{j}$ (vi) $-8\mathbf{i} + 8\mathbf{j}$
5. (i) $\begin{pmatrix} 2 \\ 1 \end{pmatrix}$ (ii) $\begin{pmatrix} -10 \\ -24 \end{pmatrix}$ (iii) $\begin{pmatrix} 0 \\ -2 \end{pmatrix}$ (iv) $\begin{pmatrix} -3 \\ 22 \end{pmatrix}$
6. (i) $\mathbf{i} + 2\mathbf{j}$, $5\mathbf{i} + \mathbf{j}$, $7\mathbf{i} + 8\mathbf{j}$ (ii) $4\mathbf{i} - \mathbf{j}$, $2\mathbf{i} + 7\mathbf{j}$, $-6\mathbf{i} - 6\mathbf{j}$
7. (i) $-3\mathbf{j}$, $2\mathbf{i} + 5\mathbf{j}$, $2\mathbf{i} + 9\mathbf{j}$ (ii) $2\mathbf{i} + 8\mathbf{j}$, $\mathbf{i} + 4\mathbf{j}$
8. $9\mathbf{i} - 10\mathbf{j}$
9. (i) $4\mathbf{i} - 5\frac{1}{2}\mathbf{j}$ (ii) $6\frac{1}{3}\mathbf{i} - 16\mathbf{j}$
10. $a = -4\frac{1}{2}$, $b = 10\frac{1}{2}$

Exercise 3C

1. (i) $113\mathbf{i} + 65\mathbf{j}$ (ii) $192\mathbf{i} - 161\mathbf{j}$ (iii) $-200\mathbf{i} - 346\mathbf{j}$ (iv) $-43\mathbf{i} + 25\mathbf{j}$
2. (i) $5.64\mathbf{i} + 2.05\mathbf{j}$ (ii) $-5.36\mathbf{i} + 4.50\mathbf{j}$ (iii) $1.93\mathbf{i} - 2.30\mathbf{j}$ (iv) $-1.45\mathbf{i} - 2.51\mathbf{j}$
3. (i) $0.65\mathbf{i} + 0.12\mathbf{j}$ (ii) $1.69\mathbf{i} - 1.18\mathbf{j}$
4. (i) 3.61, 56.3° (ii) 6.40, 51.3° (iii) 1.41, –45° (iv) 3.61, 123.7° (v) 6.40, –51.3° (vi) 6.32, –108.4° (vii) 13, 112.6° (viii) 5, 126.9°
5. 1450 m, 044°; together
6. $64.3\mathbf{i} + 76.6\mathbf{j}$, $-153.2\mathbf{i} + 128.6\mathbf{j}$, $-88.9\mathbf{i} + 205.2\mathbf{j}$
7. $-6.93\mathbf{i} - 4\mathbf{j}$, $3\mathbf{i} - 4\mathbf{j}$, $9.17\mathbf{i} - 4\mathbf{j}$
8. (ii) $-141\mathbf{i} + 141\mathbf{j}$
9. $5\mathbf{i} - \mathbf{j}$, $0.87\mathbf{i} + 3.92\mathbf{j}$, $-3.21\mathbf{i} - 4.83\mathbf{j}$, $-6\mathbf{j}$
10. (i) $35\mathbf{i} + 35\mathbf{j}$, $-87\mathbf{i} + 50\mathbf{j}$ (ii) $-51\mathbf{i} + 85\mathbf{j}$ (iii) 100 m, 329°
11. 5544 km, 051°
12. 079°, 5.10 km
13. (i) $-3\mathbf{i} + \mathbf{j} + 2\mathbf{k}$, $5\mathbf{i} - 4\mathbf{j} + 3\mathbf{k}$, $-8\mathbf{i} + 5\mathbf{j} - \mathbf{k}$ (ii) $\sqrt{6}$, $\sqrt{46}$, $\sqrt{8}$
14. (ii) (a) $\frac{4}{5}\mathbf{i} + \frac{3}{5}\mathbf{j}$ (b) $\frac{5}{13}\mathbf{i} - \frac{12}{13}\mathbf{j}$
15. (ii) (a) $\frac{1}{\sqrt{89}}(2\mathbf{i} - 6\mathbf{j} + 7\mathbf{k})$ (b) $\frac{1}{\sqrt{3}}(\mathbf{i} + \mathbf{j} + \mathbf{k})$

Exercise 3D

1. 4.42 knots
2. 329.5 kmh^{-1}, 355.1°
3. 32 ms^{-1}, 28 ms^{-1}
4. 50 to 150 ms^{-1}
5. (i) 11.8 knots, 283° (ii) 255°
6. (i) 60° (ii) 7.7 s
7. 176°, 1hr 5 mins
8. (i) 3 pm (ii) 150 km (iii) 97.6°
9. (i) 092°, 268° (ii) 5.3 km
10. (i) 18 ms^{-1} (ii) arctan $(^{18.2}/_{[50-0.3t]})$

4 MOTION IN TWO AND THREE DIMENSIONS

Exercise 4A

1. (i) $20\mathbf{i} + 34.6\mathbf{j}$ (ii) $-12\mathbf{i}$ (iii) $4.60\mathbf{i} - 3.86\mathbf{j}$ (iv) $-10\mathbf{j}$
2. $7.66\mathbf{i} + 6.43\mathbf{j}$, 5 ms^{-1} at 37° to the downward vertical
3. (i) $4t\mathbf{i} + 8\mathbf{j}$ (iv) 21.5 ms^{-1}
4. (i) $-\mathbf{i} - \mathbf{j}$ (ii) $-1^1/_3\mathbf{i} - \mathbf{j}$ (iii) $-2^2/_3\mathbf{i}$
5. $\mathbf{v} = -4\mathbf{i} - 5\mathbf{j}$, $\mathbf{a} = 0\mathbf{i} + 0\mathbf{j}$
6. 4.47, – 153.4°
7. (i) $^{t^2}/_{20}\mathbf{i} + {}^{t^3}/_{30}\mathbf{j}$ (ii) $5\mathbf{i} + {}^{100}/_3\mathbf{j}$
8. $\mathbf{v} = 2t^2\mathbf{i} + (6t - t^2)\mathbf{j}$; $\mathbf{r} = {}^2/_3t^3\mathbf{i} + (3t^2 - {}^{t^3}/_3)\mathbf{j}$
9. 8.1 ms^{-1}
10. Initial velocity = $3.54\mathbf{i} - 3.54\mathbf{j}$; $\mathbf{v} = 8.54\mathbf{i} + 11.46\mathbf{j}$; $\mathbf{s} = 52.0\mathbf{i} + 14.6\mathbf{j}$ m
12. (i) $\mathbf{v} = 15\mathbf{i} + (16 - 10t)\mathbf{j}$; $\mathbf{a} = -10\mathbf{j}$ (ii) 1.6 s (ii) 22.8 ms^{-1}
13. (ii) $20\mathbf{i} + (1 - 10t)\mathbf{j}$; $20\mathbf{i} - \mathbf{j}$ (iii) $-10\mathbf{j}$ (iv) $y = 2 + {}^x/_{20} - {}^{x^2}/_{80}$
14. (i) 50 ms^{-1}, 37° (ii) $-\mathbf{i} - \mathbf{j} - 10\mathbf{k}$ (iv) After 6 s; $222\mathbf{i} - 18\mathbf{j}$; $\mathbf{v} = 34\mathbf{i} - 6\mathbf{j} - 30\mathbf{k}$; speed = 45.7 ms^{-1}
15. (ii) 5 s (iii) 33.5 ms^{-1}, 63°
16. (i) $v \sin 35°\mathbf{i} + v \cos 35°\mathbf{j}$; $-8.66\,\mathbf{i} + 5\mathbf{j}$
    (ii) $vt \sin 35°\mathbf{i} + vt \cos 35°\mathbf{j}$; $(-8.66t + 5)\mathbf{i} + 5t\mathbf{j}$
    (iii) 6.10 kmh^{-1} (iv) 24.7 minus.

5 FORCE

Exercise 5A

1. (i) 637 N (ii) ~~104 N~~ 637N
5. (i) 11760 N (ii) 0.49 N (iii) 147 N

Exercise 5B

1. (ii) 49 N
2. (ii) 0.98 N
3. 245 N
4. 6370 N
5. 98 N
6. (ii) 19.8 N (iii) 34.3 N
7. 73.5 N, 49 N

Exercise 5C

4. (i) Weight 49 N↓, reaction from second box 49 N↑.
   (ii) Weight 49 N↓, reaction from ninth box 441 N↓, reaction from floor 490 N↑.
5. 112 N
6. 10 N, 15 N; 15 N
7. (ii) 29.4 N (iii) 29.4 N
8. 19.6 N

6 MODELLING FORCES WITH VECTORS

Exercise 6A

1. $3\mathbf{i} - 5\mathbf{j}$; 5.83 N, –59°
2. $0.196\mathbf{i} - 7\mathbf{j}$; 7.00 N, –88.4°
3. Equilibrium
4. Equilibrium
5. $9.26\mathbf{i}$; 9.26 N up incline
6. Equilibrium

Exercise 6B

1. (i) 30 N, 36.9°; 65 N, 67.4° (iii) 92.1 N, 57.9°
2. (ii) $T \cos 40°$, $T \cos 50°$ (v) 196 000 N
3. (i) 15.04 kg (ii) Both read 10 kg (iii) Both read 7.64 kg
4. (iv) (a) 9800 N, 13859 N, 9800 N (b) 9800 N, 9800 N, 9800 N
5. (iii) (a) $56.1\mathbf{i} + 61.2\mathbf{j}$ (b) 83 N, 47.5° (iv) 30.8 N at –121°
6. (ii) 4104 N (iii) 9193 N
7. (iii) 4351 N (iv) 4351N (v) No
8. (ii) 1094 N, 76 N (iii) 994 N
9. (iii) 473 N, 127 N (iv) 254 N
10. (iii) 8.69 N (iv) 1.23 kg
11. (ii) 117 N, 5.11 N (iii) 97 N forwards (iv) 3 N

7 NEWTON'S LAWS OF MOTION

Exercise 7A

1. (a)
2. (i) (d)
4. (i) 588 N
5. (i) Steady speed (ii) acceleration (iii) deceleration
6. (i) 2500 kg
7. (i) 25.6 N
8. (i) 25.3 N

Exercise 7B

1. (i) 800 N (ii) 88 500 N (iii) 0.0225 N (iv) 840 000 N (v) 8×10^{-23} N (vi) 548.8 N (vii) 0.0000875 N (viii) 10^{30} N
2. (i) 200 kg (ii) 50 kg (iii) 10 000 kg (iv) 1.02 kg
3. (i) 0.5 $\mathrm{ms^{-2}}$ (ii) 25 m
4. (i) $1\frac{2}{3}$ $\mathrm{ms^{-2}}$ (ii) 16.2 s
5. (i) 325 N (ii) 1764 N
6. (i) 2.78×10^6 N (ii) 1.1×10^7 N
7. (ii) 11300 N
8. (i) $\frac{3}{2}\mathbf{i} - \mathbf{j}$ (ii) 1.8 $\mathrm{ms^{-2}}$
9. (i) $4\mathbf{i} + 11\mathbf{j}$ (ii) $8\mathbf{i} + 8\mathbf{j}$, $2\mathbf{i} + 2\mathbf{j}$
10. (i) 13 N (ii) 90 m (iii) 13 N
11. (iii) 169.5 N
12. 13.2°
13. (i) $2t^2\mathbf{i} + 4t\mathbf{j}$ (ii) $\frac{2}{3}t^3\mathbf{i} + 2t^2\mathbf{j}$ (iii) $\frac{1}{6}t^4\mathbf{i} + \frac{2}{3}t^3\mathbf{j}$ (iv) $\mathbf{a} = 8\mathbf{i} + 8\mathbf{j}$, $\mathbf{v} = \frac{16}{3}\mathbf{i} + 8\mathbf{j}$, $\mathbf{r} = \frac{8}{3}\mathbf{i} + \frac{16}{3}\mathbf{j}$

Exercise 7C

1. (iii) is true, the rest false
2. (iv)

2. (iv) is true, the others false.
3. (i) 100 N horizontally
4. (i) 4.9 N (ii) 0 N (iii) 4.9 N
6. (i) 500 N (ii) 500 N (iii) 126.5 ms^{-1} (iv) 0.32 ms^{-1}
7. (i) 2400 N (ii) 400 N (iii) 400 N (iv) 200 N
8. (iii) $T = 5400$ N, $R_p = 540$ N, $R_L = 540$ N

Exercise 7D

1. (i) 6895, 6860, 6790, 1960 N (ii) 815 kg (iii) <9016 N
2. (ii) 1 ms^{-2} (v) 13 m
3. (iii) 1 ms^{-2}, 22 N, 27 N (iv) 5 N
4. (ii) 6.56 ms^{-2} (iii) 1.75 s (iv) 13 N
5. (i) 0.625 ms^{-1} (iii) 2.5×10^4, 1.25×10^4 N (iv) Reduces to 1×10^4 N
6. (i) 7.84 ms^{-2} (ii) The car was travelling at slightly over 30 m.p.h.
7. (ii) 0.8 ms^{-2} (iii) 40 s, $1\frac{1}{3}$ ms^{-2} (iv) 66 355 N (v) 90 kg
8. (i) 11.55 N (ii) The sledge has acceleration 1 ms^{-2}; the girl also accelerates to keep the rope taut
9. (i) 60 ms^{-1} (ii) 1.25 N (v) The lighter one
10. (ii) $11.4\mathbf{i} + 30.1\mathbf{j}$ (iii) 16.9 ms^{-2} at 21° to vertical towards Jones
11. (i) –120 ms^{-2} (ii) 120 000 N (iii) 120 000 N
12. (i) 800 N (ii) 0 N (iii) 1400 N (iv) 1400 N (v) $3T - 1000$, where T is the maximum upward pull that the builder can manage.
13. (i) It does not leave the road (just) (ii) $\mathbf{v} = 8t\mathbf{i} + \frac{1}{8}(64 - 32t + 3t^2)\mathbf{j}$; $\mathbf{a} = 8\mathbf{i} + \frac{1}{8}(-32 + 6t)\mathbf{j}$, $0 \leqslant t \leqslant 8$: $\mathbf{v} = 64\mathbf{i}$; $\mathbf{a} = 0\mathbf{i} + 0\mathbf{j}$, $8 < t \leqslant 20$ (iii) $1200\mathbf{i} + (-600 + 112.5t)\mathbf{j}$ N (iv) Because of air resistance (v) It has come to a hill.
14. (i) 3 light years (iv) $\{[0.48\mathbf{i} + 0.36\mathbf{j}], [20\mathbf{i}], [0.67\mathbf{i} - 1.33\mathbf{j} - 1.33\mathbf{k}], [-0.48\mathbf{i} - 0.64\mathbf{k}]\} \times 10^{34}\,G$ (v) $[24.67\mathbf{i} + 3.03\mathbf{j} + 0.03\mathbf{k}] \times 10^{34}\,G$ (vi) 1.85×10^{-12} ms^{-2}

8 PROJECTILES

Exercise 8A

1. (i) 17.3, 10 ms^{-1} (ii) 0, –10 ms^{-2} (iii) 34.6 m (iv) 1 s (v) 5 m
2. (i) 41, 28.7 ms^{-1} (iv) 42 m, 239.7 m
3. (i) 17.2, 8 ms^{-1} (ii) 1.645 (iii) 28.2 m (iv) 0.825 (v) 3.29 m (vi) 2.7 m, No
4. (i) 10.3, 14.7 ms^{-1} (ii) 2.91 s (iii) Into goal (iv) No
5. 45.2 s (ii) 13.55 km (iii) 535 ms^{-1} (iv) 55.9°
6. (i) 0.47 s (ii) 0.64 s (iii) 25.4 ms^{-1} (iv) 28.8 ms^{-1} (v) Too fast for a beginner. She does not serve horizontally.
7. (i) 3.14 m (ii) 0.89 s (iii) If modelled as particles, no accident; if modelled as bodies, accident.
8. (ii) 32.7 ms^{-1}
9. (i) 2.02 s (iii) 21.57 ms^{-1}
10. (iii) 57.9° (iv) +29.8 cm, –30.9 cm; the lower angle is slightly more accurate.
11. (i) 26.1 ms^{-1} (ii) 27.35 ms^{-1} (iii) $26.15 < u < 26.88$
12. 25.5 m

Exercise 8B

2. (i) $y = \frac{3}{4}x - \frac{x^2}{320}$
3. (ii) $u > 9.05$ ms^{-1} (iii) $u < 8.71$ ms^{-1}
4. (ii) 24.1 m
5. (i) $y = x\tan\alpha - \dfrac{gx^2}{5000x^2\alpha}$

   (iii) (153, 76.5) (iv) 171 (v) 127.5
6. (ii) 1.8 or 0.6 (iii) 22.5 m when $\tan\alpha = 1.2$ (iv) 15.4 m

Index